W0114853

SAMPLE SIZE CHOICE

STATISTICS: Textbooks and Monographs

A Series Edited by

D. B. Owen, Coordinating Editor
Department of Statistics
Southern Methodist University
Dallas, Texas

R. G. Cornell, Associate Editor
for Biostatistics
University of Michigan

W. J. Kennedy, Associate Editor
for Statistical Computing
Iowa State University

A. M. Kshirsagar, Associate Editor
for Multivariate Analysis and
Experimental Design
University of Michigan

E. G. Schilling, Associate Editor
for Statistical Quality Control
Rochester Institute of Technology

Vol. 1: The Generalized Jackknife Statistic, *H. L. Gray and W. R. Schucany*
Vol. 2: Multivariate Analysis, *Anant M. Kshirsagar*
Vol. 3: Statistics and Society, *Walter T. Federer*
Vol. 4: Multivariate Analysis: A Selected and Abstracted Bibliography, 1957-1972, *Kocherlakota Subrahmaniam and Kathleen Subrahmaniam* (out of print)
Vol. 5: Design of Experiments: A Realistic Approach, *Virgil L. Anderson and Robert A. McLean*
Vol. 6: Statistical and Mathematical Aspects of Pollution Problems, *John W. Pratt*
Vol. 7: Introduction to Probability and Statistics (in two parts), Part I: Probability; Part II: Statistics, *Narayan C. Giri*
Vol. 8: Statistical Theory of the Analysis of Experimental Designs, *J. Ogawa*
Vol. 9: Statistical Techniques in Simulation (in two parts), *Jack P. C. Kleijnen*
Vol. 10: Data Quality Control and Editing, *Joseph I. Naus* (out of print)
Vol. 11: Cost of Living Index Numbers: Practice, Precision, and Theory, *Kali S. Banerjee*
Vol. 12: Weighing Designs: For Chemistry, Medicine, Economics, Operations Research, Statistics, *Kali S. Banerjee*
Vol. 13: The Search for Oil: Some Statistical Methods and Techniques, *edited by D. B. Owen*
Vol. 14: Sample Size Choice: Charts for Experiments with Linear Models, *Robert E. Odeh and Martin Fox*
Vol. 15: Statistical Methods for Engineers and Scientists, *Robert M. Bethea, Benjamin S. Duran, and Thomas L. Boullion*
Vol. 16: Statistical Quality Control Methods, *Irving W. Burr*
Vol. 17: On the History of Statistics and Probability, *edited by D. B. Owen*
Vol. 18: Econometrics, *Peter Schmidt*
Vol. 19: Sufficient Statistics: Selected Contributions, *Vasant S. Huzurbazar (edited by Anant M. Kshirsagar)*
Vol. 20: Handbook of Statistical Distributions, *Jagdish K. Patel, C. H. Kapadia, and D. B. Owen*
Vol. 21: Case Studies in Sample Design, *A. C. Rosander*
Vol. 22: Pocket Book of Statistical Tables, *compiled by R. E. Odeh, D. B. Owen, Z. W. Birnbaum, and L. Fisher*

Vol. 23: The Information in Contingency Tables, *D. V. Gokhale and Solomon Kullback*
Vol. 24: Statistical Analysis of Reliability and Life-Testing Models: Theory and Methods, *Lee J. Bain*
Vol. 25: Elementary Statistical Quality Control, *Irving W. Burr*
Vol. 26: An Introduction to Probability and Statistics Using BASIC, *Richard A. Groeneveld*
Vol. 27: Basic Applied Statistics, *B. L. Raktoe and J. J. Hubert*
Vol. 28: A Primer in Probability, *Kathleen Subrahmaniam*
Vol. 29: Random Processes: A First Look, *R. Syski*
Vol. 30: Regression Methods: A Tool for Data Analysis, *Rudolf J. Freund and Paul D. Minton*
Vol. 31: Randomization Tests, *Eugene S. Edgington*
Vol. 32: Tables for Normal Tolerance Limits, Sampling Plans, and Screening, *Robert E. Odeh and D. B. Owen*
Vol. 33: Statistical Computing, *William J. Kennedy, Jr. and James E. Gentle*
Vol. 34: Regression Analysis and Its Application: A Data-Oriented Approach, *Richard F. Gunst and Robert L. Mason*
Vol. 35: Scientific Strategies to Save Your Life, *I. D. J. Bross*
Vol. 36: Statistics in the Pharmaceutical Industry, *edited by C. Ralph Buncher and Jia-Yeong Tsay*
Vol. 37: Sampling from a Finite Population, *J. Hajek*
Vol. 38: Statistical Modeling Techniques, *S. S. Shapiro*
Vol. 39: Statistical Theory and Inference in Research, *T. A. Bancroft and C.-P. Han*
Vol. 40: Handbook of the Normal Distribution, *Jagdish K. Patel and Campbell B. Read*
Vol. 41: Recent Advances in Regression Methods, *Hrishikesh D. Vinod and Aman Ullah*
Vol. 42: Acceptance Sampling in Quality Control, *Edward G. Schilling*
Vol. 43: The Randomized Clinical Trial and Therapeutic Decisions, *edited by Niels Tygstrup, John M. Lachin, and Erik Juhl*
Vol. 44: Regression Analysis of Survival Data in Cancer Chemotherapy, *Walter H. Carter, Jr., Galen L. Wampler, and Donald M. Stablein*
Vol. 45: A Course in Linear Models, *Anant M. Kshirsagar*
Vol. 46: Clinical Trials: Issues and Approaches, *edited by Stanley H. Shapiro and Thomas H. Louis*
Vol. 47: Statistical Analysis of DNA Sequence Data, *edited by B. S. Weir*
Vol. 48: Nonlinear Regression Modeling: A Unified Practical Approach, *David A. Ratkowsky*
Vol. 49: Attribute Sampling Plans, Tables of Tests and Confidence Limits for Proportions, *Robert E. Odeh and D. B. Owen*
Vol. 50: Experimental Design, Statistical Models, and Genetic Statistics, *edited by Klaus Hinkelmann*
Vol. 51: Statistical Methods for Cancer Studies, *edited by Richard G. Cornell*
Vol. 52: Practical Statistical Sampling for Auditors, *Arthur J. Wilburn*
Vol. 53: Statistical Signal Processing, *edited by Edward J. Wegman and James G. Smith*
Vol. 54: Self-Organizing Methods in Modeling: GMDH Type Algorithms, *edited by Stanley J. Farlow*
Vol. 55: Applied Factorial and Fractional Designs, *Robert A. McLean and Virgil L. Anderson*
Vol. 56: Design of Experiments: Ranking and Selection, *edited by Thomas J. Santner and Ajit C. Tamhane*
Vol. 57: Statistical Methods for Engineers and Scientists. Second Edition, Revised and Expanded, *Robert M. Bethea, Benjamin S. Duran, and Thomas L. Boullion*
Vol. 58: Ensemble Modeling: Inference from Small-Scale Properties to Large-Scale Systems, *Alan E. Gelfand and Crayton C. Walker*

Vol. 59: Computer Modeling for Business and Industry, *Bruce L. Bowerman and Richard T. O'Connell*
Vol. 60: Bayesian Analysis of Linear Models, *Lyle D. Broemeling*
Vol. 61: Methodological Issues for Health Care Surveys, *Brenda Cox and Steven Cohen*
Vol. 62: Applied Regression Analysis and Experimental Design, *Richard J. Brook and Gregory C. Arnold*
Vol. 63: Statpal: A Statistical Package for Microcomputers – PC-DOS Version for the IBM PC and Compatibles, *Bruce J. Chalmer and David G. Whitmore*
Vol. 64: Statpal: A Statistical Package for Microcomputers – Apple Version for the II, II+, and IIe, *David G. Whitmore and Bruce J. Chalmer*
Vol. 65: Nonparametric Statistical Inference, Second Edition, Revised and Expanded, *Jean Dickinson Gibbons*
Vol. 66: Design and Analysis of Experiments, *Roger G. Petersen*
Vol. 67: Statistical Methods for Pharmaceutical Research Planning, *Sten W. Bergman and John C. Gittins*
Vol. 68: Goodness-of-Fit Techniques, *edited by Ralph B. D'Agostino and Michael A. Stephens*
Vol. 69: Statistical Methods in Discrimination Litigation, *edited by D. H. Kaye and Mikel Aickin*
Vol. 70: Truncated and Censored Samples from Normal Populations, *Helmut Schneider*
Vol. 71: Robust Inference, *M. L. Tiku, W. Y. Tan, and N. Balakrishnan*
Vol. 72: Statistical Image Processing and Graphics, *edited by Edward J. Wegman and Douglas J. DePriest*
Vol. 73: Assignment Methods in Combinatorial Data Analysis, *Lawrence J. Hubert*
Vol. 74: Econometrics and Structural Change, *Lyle D. Broemeling and Hiroki Tsurumi*
Vol. 75: Multivariate Interpretation of Clinical Laboratory Data, *Adelin Albert and Eugene K. Harris*
Vol. 76: Statistical Tools for Simulation Practitioners, *Jack P. C. Kleijnen*
Vol. 77: Randomization Tests, Second Edition, *Eugene S. Edgington*
Vol. 78: A Folio of Distributions: A Collection of Theoretical Quantile-Quantile Plots, *Edward B. Fowlkes*
Vol. 79: Applied Categorical Data Analysis, *Daniel H. Freeman, Jr.*
Vol. 80: Seemingly Unrelated Regression Equations Models : Estimation and Inference, *Virendra K. Srivastava and David E. A. Giles*
Vol. 81: Response Surfaces: Designs and Analyses, *Andre I. Khuri and John A. Cornell*
Vol. 82: Nonlinear Parameter Estimation: An Integrated System in BASIC, *John C. Nash and Mary Walker-Smith*
Vol. 83: Cancer Modeling, *edited by James R. Thompson and Barry W. Brown*
Vol. 84: Mixture Models: Inference and Applications to Clustering, *Geoffrey J. McLachlan and Kaye E. Basford*
Vol. 85: Randomized Response: Theory and Techniques, *Arijit Chaudhuri and Rahul Mukerjee*
Vol. 86: Biopharmaceutical Statistics for Drug Development, *edited by Karl E. Peace*
Vol. 87: Parts per Million Values for Estimating Quality Levels, *Robert E. Odeh and D. B. Owen*
Vol. 88: Lognormal Distributions: Theory and Applications, *edited by Edwin L. Crow and Kunio Shimizu*
Vol. 89: Properties of Estimators for the Gamma Distribution, *K. O. Bowman and L. R. Shenton*
Vol. 90: Spline Smoothing and Nonparametric Regression, *Randall L. Eubank*
Vol. 91: Linear Least Squares Computations, *R. W. Farebrother*
Vol. 92: Exploring Statistics, *Damaraju Raghavarao*

Vol. 93: Applied Time Series Analysis for Business and Economic Forecasting, *Sufi M. Nazem*
Vol. 94: Bayesian Analysis of Time Series and Dynamic Models, *edited by James C. Spall*
Vol. 95: The Inverse Gaussian Distribution: Theory, Methodology, and Applications, *Raj S. Chhikara and J. Leroy Folks*
Vol. 96: Parameter Estimation in Reliability and Life Span Models, *A. Clifford Cohen and Betty Jones Whitten*
Vol. 97: Pooled Cross-Sectional and Time Series Data Analysis, *Terry E. Dielman*
Vol. 98: Random Processes: A First Look, Second Edition, Revised and Expanded, *R. Syski*
Vol. 99: Generalized Poisson Distributions: Properties and Applications, *P.C. Consul*
Vol. 100: Nonlinear L_p-Norm Estimation, *René Gonin and Arthur H. Money*
Vol. 101: Model Discrimination for Nonlinear Regression Models, *Dale S. Borowiak*
Vol. 102: Applied Regression Analysis in Econometrics, *Howard E. Doran*
Vol. 103: Continued Fractions in Statistical Applications, *K.O. Bowman and L.R. Shenton*
Vol. 104: Statistical Methodology in the Pharmaceutical Sciences, *Donald A. Berry*
Vol. 105: Experimental Design in Biotechnology, *Perry D. Haaland*
Vol. 106: Statistical Issues in Drug Research and Development, *edited by Karl E. Peace*
Vol. 107: Handbook of Nonlinear Regression Models, *David A. Ratkowsky*
Vol. 108: Robust Regression: Analysis and Applications, *edited by Kenneth D. Lawrence and Jeffrey L. Arthur*
Vol. 109: Statistical Design and Analysis of Industrial Experiments, *edited by Subir Ghosh*
Vol. 110: *U*-Statistics: Theory and Practice, *A. J. Lee*
Vol. 111: A Primer in Probability, Second Edition, Revised and Expanded, *Kathleen Subrahmaniam*
Vol. 112: Data Quality Control: Theory and Pragmatics, *edited by Gunar E. Liepins and V. R. R. Uppuluri*
Vol. 113: Engineering Quality by Design: Interpreting the Taguchi Approach, *Thomas B. Barker*
Vol. 114: Survivorship Analysis for Clinical Studies, *Eugene K. Harris and Adelin Albert*
Vol. 115: Statistical Analysis of Reliability and Life-Testing Models, Second Edition, *Lee J. Bain and Max Engelhardt*
Vol. 116: Stochastic Models of Carcinogenesis, *Wai-Yuan Tan*
Vol. 117: Statistics and Society: Data Collection and Interpretation, Second Edition, Revised and Expanded, *Walter T. Federer*
Vol. 118: Handbook of Sequential Analysis, *B. K. Ghosh and P. K. Sen*
Vol. 119: Truncated and Censored Samples: Theory and Applications, *A. Clifford Cohen*
Vol. 120: Survey Sampling Principles, *E. K. Foreman*
Vol. 121: Applied Engineering Statistics, *Robert M. Bethea and R. Russell Rinehart*
Vol. 122: Sample Size Choice: Charts for Experiments with Linear Models, Second Edition, *Robert E. Odeh and Martin Fox*

ADDITIONAL VOLUMES IN PREPARATION

SAMPLE SIZE CHOICE

Charts for Experiments with Linear Models

Second Edition

Robert E. Odeh
University of Victoria
Victoria, British Columbia, Canada

Martin Fox
Michigan State University
East Lansing, Michigan

CRC Press is an imprint of the
Taylor & Francis Group, an **informa** business

First published 1991 by Marcel Dekker

Published 2020 by CRC Press
Taylor & Francis Group
6000 Broken Sound Parkway NW, Suite 300
Boca Raton, FL 33487-2742

ISBN 13: 978-0-8247-8600-7 (hbk)

Visit the Taylor & Francis Web site at
http://www.taylorandfrancis.com

and the CRC Press Web site at
http://www.crcpress.com

Library of Congress Cataloging-in-Publication Data

Odeh, Robert E.
Sample size choice: charts for experiments with linear models / Robert E. Odeh, Martin Fox. -- 2nd ed.
p. cm. -- (Statistics, textbooks and monographs; v. 122)
Includes bibliographical references and index.
ISBN 0-8247-8600-9
1. Experimental design--Charts, diagrams, etc. 2. Statistical hypothesis testing--Charts, diagrams, etc. 3. Sampling (Statistics)--Charts, diagrams, etc. 4. Linear models (Statistics)--Charts, diagrams, etc. I. Fox, Martin. II. Title. III. Series.
QA279.033 1991
519.5--dc20 91-22887
CIP

To the memory of Jerzy Neyman and Egon S. Pearson, who together developed our modern theory of testing of statistical hypotheses and led us to consideration of power of tests in the design of experiments and analysis of data.

PREFACE TO THE SECOND EDITION

In the second edition the material has been rearranged so as to bring the reader more expeditiously to the examples (now Chapter 2). The introduction has been limited to the power of the F−test and a description of the charts. The remainder of the theoretical discussion is deferred to Chapter 3 and is followed by Chapter 4 (formerly Chapter 2) which is a survey of the literature. The last two chapters are as in the first edition.

We have extensively rewritten the textual material with, we believe, greater clarity. Errors have be corrected, including gross errors in the section on regression. Examples have been added to demonstrate the choice of sample size for:

1. The one-way layout when the null hypothesis states that $k < I$ out of the I means are equal (Example 2.2.4). Also considered are unequal sample sizes for this case (Example 2.2.5).

2. Latin squares (Example 2.5.1).

3. Profile analysis (Example 2.7.3).

References have been updated and improved computational algorithms, not available to us for the first edition, are mentioned.

The authors wish to thank Charles E. Cress, John Gill, John Hall, and Carl Ramm who read the first edition and suggested some of the changes and additions mentioned above.

ROBERT E. ODEH

MARTIN FOX

PREFACE TO THE FIRST EDITION

When designing experiments intended for testing statistical hypotheses, one should consider both the desired level of significance (usually denoted by α) and the desired power of the test. Generally, one can control both of these quantities by selection of the number n of replicates, the power with fixed α increasing as n increases. These charts are intended to enable one to find n achieving a given α and power in experiments for which linear models are appropriate. A wide range of both quantities is provided.

Prior to the advent of the high-speed computer, tables of statistical functions were very limited in their coverage of entry values. Traditionally, and for no other good reason, this limitation restricted tables used for hypothesis testing to levels of significance .01 and .05. By now, some degree of magic has been associated with these two levels. With the general availability of high-speed computation, the range of levels can by extended and research workers can use the level most appropriate to their own work.

The authors wish to thank the editors of the Annals of Mathematical Statistics for permission to use the format of charts introduced in their pages by Fox (1956). This collection is an expansion of those contained in that journal.

We also wish to thank Joseph L. Hodges, Jr. who suggested the format of the Fox (1956) charts of which these are an extension.

Parts of the manuscript, in several stages of readiness, were read by Kenneth J. Arnold, Charles Cress, John L. Gill, William C. Guenther, W. Keith Hastings, Donald B. Owen, James Stapleton, and M.L. Tiku. Extensive rewriting resulted from their comments. Any remaining difficulties are our own fault.

Without Noralee Burkhart's meticulous work, the manuscript could have never appeared in print. We thank her for typing the preliminary drafts as well as the final version.

The values for the charts and tables were computed on the IBM/145 at the University of Victoria. We are particularly grateful to Eric Gelling, who assisted with much of the computer programming. We also wish to express our appreciation to the University of Victoria Computing Centre for their generous assistance.

The original charts were produced on a Calcomp Plotter at the University of Victoria. The charts were redrawn, for publication, by the Graphics Division, Technical Media Center, Michigan State University. We are particularly indebted to Mr. D.J. Wilkening and his staff.

Amanda Linnell proofread the final copy of the manuscript and uncovered many errors and inconsistencies.

ROBERT E. ODEH

MARTIN FOX

CONTENTS

SAMPLE SIZE CHOICE

PART ONE

Chapter 1

INTRODUCTION

We assume that users of these charts are familiar with the Neyman-Pearson theory of hypothesis testing including the notion of power. We also assume familiarity with the general linear hypothesis (multiple regression) problem as well as the special case of analysis of variance.

The remainder of this chapter consists of discussions of the power of F-tests and the ranges of parameter values covered by the charts.

1.1 Power of the F-test

When the F-test is used in general linear hypothesis (e.g, analysis of variance, regression), the power of the test or, equivalently, the operating characteristic (1-power), is a function of a quantity known as the noncentrality parameter. As a first step in defining the noncentrality parameter, we give three equivalent computations of a quantity denoted by S^*. The numerator and denominator degrees of freedom are denoted by f_1 and f_2, respectively.

1. Replace each observation in the numerator sum of squares in the F-statistic by its expected value.

2. The numerator *E(MS)* is $\sigma^2 + S^*/f_1$

3. Express the problem in canonical form. The observations are Y_i $(i = 1, \ldots, N)$, which are independent random variables, normally distributed with means μ_i $(i = 1, \ldots, N)$ and common variance σ^2 where $\mu_1 = \ldots = \mu_{f_2} = 0$.

The hypothesis is that $\mu_{f_2+1} = \cdots = \mu_{f_2+f_1} = 0$. Then,

$$S^* = \sum_{i=f_2+1}^{f_2+f_1} \mu_i^2 \tag{1.1}$$

The noncentrality parameter is

$$\varphi = \sqrt{\frac{S^*}{\sigma^2(f_1+1)}} \tag{1.2}$$

Fix the level of significance (α) and the power and consider the pairs (f_1, f_2) for which a given value of φ will yield the fixed power. Plot the curves along which φ is constant as f_1 and f_2 vary. If this is done using reciprocal scales for f_1 and f_2, then these curves are very nearly straight lines making visual interpolation very easy. Charts of this form are useful for obtaining the sample size needed to achieve the desired power. Using these observations, one of the authors (Fox, 1956) presented charts for values of α =.01, .05 and power (denoted by β) equal to .5, .7, .8, and .9. These charts have been reprinted by Scheffé (1959) and Owen (1962).

The Fox charts are extended here.

1.2 Range of the Charts

In designing experiments for use in testing statistical hypotheses, the trade-off between making α small, making power large, and the cost of experimentation needs to be considered. Thus, charts or tables covering an extensive range of values of both α and power are needed. The charts given here satisfy this need and may be used in connection with any linear model.

Before the advent of the high-speed computer the sheer labor of computation severely limited the range of parameters over which tables and charts could be computed, hence limiting the range of possible choices of, for example, α and power. Such limitations are no longer a serious factor and the experimenter should be much more flexible in designing experiments. Thus, in many cases, value of α other than .01 and .05 (those most commonly used) should be considered.

The charts are given for $f_1 \geq 1$, $f_2 \geq 4$, and the pairs (α, power) as listed in Table 1.1.

A table of critical values for the F-distribution, whose α-entries are those values given in Table 1.1 is included as are similar tables of the χ^2– and t-distribution. Also included is a discussion of the computational procedure.

α	Power
.001	.005, .01, .025, .05, .1, .2, .3, .4, .5, .6, .7, .8, .9, .95, .975, .99, .995
.005	.01, .025, .05, .1, .2, .3, .4, .5, .6, .7, .8, .9, .95, .975, .99, .995
.01	.025, .05, .1, .2, .3, .4, .5, .6, .7, .8, .9, .95, .975, .99, .995
.025	.05, .1, .2, .3, .4, .5, .6, .7, .8, .9, .95, .975, .99, .995
.05	.1, .2, .3, .4, .5, .6, .7, .8, .9, .95, .975, .99, .995
.10	.2, .3, .4, .5, .6, .7, .8, .9, .95, .975, .99, .995
.25	.3, .4, .5, .6, .7, .8, .9, .95, .975, .99, .995
.50	.6, .7, .8, .9, .95, .975, .99, .995

Table 1.1

Chapter 2

EXAMPLES

2.1 General Remarks

Each of the sections that follow contain several examples of the use of the charts in Part Three for a particular form of the linear model (e.g., Sec. 2.2 contains examples for the one-way layout). These examples are actually culled from experimental literature.

In using these charts the steps to be followed are:

1. The model is stated.

2. The null hypothesis is given in terms of the parameters of the model. This is usually stated in the form H_0: *statement about the parameters.*

3. The value of α is fixed.

4. An alternative hypothesis is given in terms of the parameters of the model. This is also usually stated in the form H_1: *statement about the parameters.*

5. The value of the power desired for H_1 is fixed.

6. For the value of f_1 determined by H_0 and $f_2 = \infty$ read φ from the appropriate charts (determined in steps 3 and 5).

7. Use the value of φ obtained in step 6 and (1.2) to solve for n, rounding up to the next integer (S^* and, hence, φ will always depend on n).

8. Using the latest value of n, compute φ and f_2.

9. From the chart, find the intersection of the curve for the value of φ in step 8 with the line for f_1 in step 6 and read the value of f_2 at that point.

10. If the value of f_2 in step 8 is less than that in step 9, increase n by 1 and return to step 8; otherwise the current value of n is the desired value.

The values of f_2 and φ obtained in step 8 are the correct values for the value of n being tried at that point. The value of f_2 obtained in step 9 is the value which, with the above value of φ , would guarantee the desired power. Increasing n increases S^* and, hence, φ as well as the value of f_2 in step 8. The increase in φ results in a decrease in f_2 obtained in step 9 (see the charts). Thus, this procedure will eventually end and yield the smallest value of n for which the preassigned power is obtained.

Steps 1 to 10 will be illustrated through examples. The reader may disagree with our interpolated values of φ as read from the charts. Note, however, that in most cases in which f_2 is small or moderate (up to about $f_2 = 20$), small errors in reading φ will lead to no changes in the values of n obtained. For larger values of f_2, the errors will generally be small as a percentage of the correct n.

If quantities z_{ij} $(i = 1,\ldots,m;\ j = 1,\ldots,n)$ are given, the following notation will be used for sums and means:

$$z_{i.} = \sum_{j=1}^{n} z_{ij} \qquad \bar{z}_{i.} = z_{i.}/n$$

$$z_{.j} = \sum_{i=1}^{m} z_{ij} \qquad \bar{z}_{.j} = z_{.j}/m$$

$$z_{..} = \sum_{i=1}^{m}\sum_{j=1}^{n} z_{ij} \qquad \bar{z}_{..} = z_{..}/(mn)$$

This notation will be used with constants and random variables and can be generalized to any number of subscripts.

2.2 One-way Layout

Steps 1 and 2 will be carried out in general.

1. Assume that n observations will be taken in each cell. The model is usually considered in the form

$$Y_{ij} = \mu_i + \epsilon_{ij} \; (i = 1, \ldots, I;\; j = 1, \ldots, n) \tag{2.1}$$

where the μ_i are fixed but unknown numbers and the ϵ_{ij} are random fluctuations which are independent and normally distributed with mean 0 and variance σ^2. Here I is the number of cells. Hence, $N = In$ in the notation of Chap. 1. It is convenient to rewrite (2.1) in the form

$$Y_{ij} = \overline{\mu}_. + \delta_i + \epsilon_{ij} \tag{2.2}$$

To obtain (2.2), set $\delta_i = \mu_i - \overline{\mu}_.$ so that δ_i represents the effect of cell i. It is easy to see that $\delta_. = 0$.

2. H_0: $\delta_i = 0 \; (i = 1, \ldots, I)$ or equivalently, H_0: $\mu_1 = \mu_2 = \cdots = \mu_I$. Set

$$SSTr = n\sum_i (\overline{Y}_{i.} - \overline{Y}_{..})^2 = \sum_i Y_{i.}^2/n - Y_{..}^2/(In)$$

$$SST = \sum_i \sum_j (Y_{ij} - \overline{Y}_{..})^2 = \sum_i \sum_j Y_{ij}^2 - Y_{..}^2/(In)$$

$$SSE = SST - SSTr$$

The test statistic is $MSTr/MSE$. The second expressions for $SSTr$ and SST are the most useful for calculations.

Since $E(\overline{Y}_{i.}) = \overline{\mu}_. + \delta_i$ and $E(\overline{Y}_{..}) = \overline{\mu}_.$, the first expression for $SSTr$ results in $S^* = n\sum_i \delta_i^2$. Hence,

$$\varphi = \sqrt{n\sum_i \delta_i^2 / I\sigma^2} \tag{2.3}$$

Table 2.1 is the *ANOVA* table for this model.

Source	SS	$d.f.$	$E(MS)$
Treatments	$SSTr$	$I - 1$	$\sigma^2 + n\sum_i \delta_i^2/(I-1)$
Error	SSE	$I(n-1)$	σ^2
Total	SST	$nI - 1$	

Table 2.1

Note the relationship between S^* and $E(MSTr)$. In Example 2.2.2 the *ANOVA* table will be computed to illustrate these calculations.

EXAMPLE 2.2.1. (Steer Feeding). Hardison and Reid (1953) measured intakes of dry matter for hand-fed and grazing steers. The null hypothesis is that the two feeding methods result in the same dry matter intakes. The appropriate model is (2.2) with $I = 2$. Thus, $\delta_2 = -\delta_1$ and (2.3) yields $\varphi = \sqrt{n\delta_1^2/\sigma^2}$

Now proceed through steps 3 to 10:

3. Set $\alpha = .05$

4, 5. Set the power at .8 when $\delta_1^2 = 2\sigma^2$

6. Since $I = 2$, it follows that $f_1 = 1$. Reading the chart for $\alpha = .05$ and power .8, for $f_1 = 1$, $f_2 = \infty$, it is required that $\varphi = 1.98$

7. For $\delta_1^2 = 2\sigma^2$,

$$\varphi = \sqrt{2n} \tag{2.4}$$

Solving (2.4) for n and rounding to the next larger integer yields $n = 2$

8-10. Table 2.2 summarizes the remaining steps of the iteration. Thus, $n = 4$ suffices. In fact, Hardison and Reid used $n = 3$. The critical value for $n = 4$ is $F_{.05}(1, 6) = 5.987$ (See Table 1.4 in Part Two).

See Secs. 2.8 and 2.9 for alternative treatments of this example.

n	$\varphi = \sqrt{2n}$	$f_2 = 2(n-1)$	f_2 from chart
2	2.00	2	100
3	2.45	4	6
4	2.83	6	<4

Table 2.2

EXAMPLE 2.2.2. (Relation between archery scores and ordering of shots). The standard archery round consists of six shots at each of the ranges 30, 40, and 50 yards, in that order. Schroeder (1945) (also found in Walker and Lev, 1953) gave six archery lessons to each member of each of three groups of equal size. For each lesson, the order of ranges was different. For each group the ordering of the lessons differed.

Since $I = 3$, it follows that $f_1 = 2$ and $f_2 = 3(n-1)$, where n is the number of individuals in each group. The hypothesis to be tested is that there is no difference in 50 yard scores between groups. Set $\alpha = .025$ and let the power be .7 when $\sum_i \delta_i^2/\sigma^2 = 1$. Then, (2.3) yields $\varphi = \sqrt{n/3}$.

Initially $n = 10$ and the iteration is displayed in Table 2.3.

n	$\varphi = \sqrt{n/3}$	$f_2 = 3(n-1)$	f_2 from chart
10	1.82	27	∞
11	1.91	30	30

Table 2.3

In fact, Schroeder set $n = 11$. The critical value for $n = 11$ is $F_{.025}(2, 30) = 4.182$

The *ANOVA* table (Table 2.1) will be computed below. The data and sums are given in Table 2.4. Also

$$\sum_{i=1}^{3}\sum_{j=1}^{11} Y_{ij}^2 = 373,977$$

j	$i=1$	$i=2$	$i=3$	
1	99	104	41	
2	114	26	83	
3	51	136	61	
4	78	87	189	
5	134	187	88	
6	71	34	80	
7	66	106	141	
8	33	63	112	
9	146	155	141	
10	80	68	112	
11	105	29	173	
$Y_{i.}$	977	995	1,221	$Y_{..}$ =3,193

Table 2.4

Then,

$$\frac{\sum_{i=1}^{3} Y_{i.}^2}{11} = \frac{(977)^2 + (995)^2 + (1221)^2}{11} = 312,309$$

and

$$\frac{Y_{..}^2}{33} = \frac{(3193)^2}{33} = 308,947$$

Source	SS	$d.f.$	MS	$E(MS)$
Treatments	3,362	2	1,681	$\sigma^2 + \frac{11}{2}(\delta_1^2 + \delta_2^2 + \delta_3^2)$
Error	61,688	30	2,056	σ^2
Total	65,030	32		

Table 2.5

The computed *ANOVA* table is given in Table 2.5. The value of the test statistic is $1681/2056 = .818$. Since this is less than the critical value 4.182, H_0 is accepted. Had the value of the test statistic been greater than 4.182, H_0 would have been rejected.

EXAMPLE 2.2.3 (Effect on red clover nitrogen content of inoculation with certain cultures). Erdman (1946) (data also found in Steel and Torrie, 1960) inoculated samples of red clover with a culture containing material from five red clover and five alfalfa strains. Consideration will be restricted to part of the data in which six groups were tested, one each with a mix containing one of the red alfalfa strains and one with a blend of all five. Each mix considered contained all of the alfalfa strains. Measurements were made later of the nitrogen content of the plants.

Since there are six treatments, $I = 6$, so that $f_1 = 5$ and $f_2 = 6(n-1)$. The hypothesis tested is that there is no difference in nitrogen content between groups. Suppose that Erdman had set $\alpha = .01$ and wished the power to be equal to .975 when $\sum_i \delta_i^2/\sigma^2 = 4$. Then, $\varphi = \sqrt{2n/3}$. The iteration is displayed in Table 2.6.

n	$\varphi = \sqrt{2n/3}$	$f_2 = 6(n-1)$	f_2 from chart
8	2.31	42	200
9	2.45	48	35

Table 2.6

In fact, Erdman took $n = 5$. By interpolation the critical value for $n = 9$ is $F_{.01}(5,48) = 3.425$. Interpolation would only be required if the F-statistic is between $F_{.01}(5,50) = 3.408$ and $F_{.01}(5,40) = 3.514$

Another interesting problem is to test the hypothesis H_0: $\mu_1 = \cdots = \mu_k$ where $k < I$. For this case SSE and f_2 are unchanged while $f_1 = k - 1$. Set $W = \sum_{i=1}^k \overline{Y}_{i\cdot}/k$, the estimate of the common value of $\mu_1, \ldots, \mu_k$ under H_0. Then the numerator sum of squares is $SSH = n\sum_{i=1}^k(\overline{Y}_{i\cdot} - W)^2$. Since $E(\overline{Y}_{i\cdot}) = \mu_i$ and $E(W) = \sum_{i=1}^k \mu_i/k$ (denoted here by ν), it follows that $S^* = n\sum_{i=1}^k(\mu_i - \nu)^2$.

In most common applications $k = 2$ so that $f_1 = 1, S^* = n(\mu_1 - \mu_2)^2$ and

$$\varphi = \sqrt{\frac{n(\mu_1 - \mu_2)^2}{2\sigma^2}} \tag{2.4}$$

EXAMPLE 2.2.4 (See example 2.2.2). Suppose that only the scores for groups A and B are to be compared. Set $\alpha = .025$ and let the power be .7 when $(\mu_1 - \mu_2)^2 = \sigma^2$. Then, from (2.4) we have $\varphi = \sqrt{n/2}$. The iteration is displayed in Table 2.7.

n	$\varphi = \sqrt{n/2}$	$f_2 = 3(n-1)$	f_2 from chart
8	2.00	21	50
9	2.12	24	16

Table 2.7

The critical value is $F_{.025}(1, 24) = 5.717$

It would be reasonable in the problem just considered to use larger sample sizes for the first k cells than in the remaining cells. Let n_i be the number of observations in cell i, and

$$W = \sum_{i=1}^{k}\sum_{j=1}^{n_i} Y_{ij} / \sum_{i=1}^{k} n_i = \sum_{i=1}^{k} n_i \overline{Y}_{i.} / \sum_{i=1}^{k} n_i$$

Then, $SSH = \sum_{j=1}^{k} n_i(\overline{Y}_{i.} - W)^2$, $E(\overline{Y}_{i.}) = \mu_i$ and $E(W) = \sum_{i=1}^{k} n_i\mu_i / \sum_{i=1}^{k} n_i$ (denoted here by ν). Hence, $S^* = \sum_{i=1}^{k} n_i(\mu_i - \nu)^2$. Also $f_2 = \sum_{i=1}^{I} n_i - I$. For $k = 2$ with $n_1 = n_2$ ($= m$, say), it follows that $S^* = m(\mu_1 - \mu_2)^2$ and

$$\varphi = \sqrt{\frac{m(\mu_1 - \mu_2)^2}{2\sigma^2}} \tag{2.5}$$

Furthermore, $f_2 = 2m + n - 3$.

EXAMPLE 2.2.5 (See examples 2.2.2 and 2.2.4). Let the requirement be as in Example 2.2.4 and assume $m = 2n$ is required. Then $f_1 = 1$, $f_2 = 5n - 3$ and $\varphi = \sqrt{n}$. The iteration is displayed in Table 2.8.

n	$\varphi = \sqrt{n}$	$f_2 = 5n - 3$	f_2 from chart
4	2.00	17	50
5	2.24	22	10

Table 2.8

In this case $n_1 = n_2 = 10$ and $n_3 = 5$ so that $N = 25$, while in Example 2.2.4, $N = 27$. The critical value is $F_{.025}(1, 22) = 5.786$

2.3 Two-way Layout

Suppose that there are two factors influencing the outcome of a trial instead of only one. The model is

$$Y_{ijk} = \mu + \delta_i + \gamma_j + \rho_{ij} + \epsilon_{ijk} \tag{2.6}$$

where Y_{ijk} is the *k-th* observation obtained at the *i-th* level of one factor and the *j-th* level of the other and the ϵ_{ijk} are random fluctuations which are independent and normally distributed with means 0 and unknown variances σ^2. The remaining terms in (2.6) are unknown parameters which, without loss of generality, may be assumed to satisfy

$$\delta_. = \gamma_. = \rho_{.j} = \rho_{i.} = 0. \tag{2.7}$$

With (2.7) imposed, the δ_i and γ_j are called the *main effects* of the respective factors and the ρ_{ij} are the *interactions*.

Assume that the ranges of the subscripts are $i = 1, \ldots, I$; $j = 1, \ldots, J$; and $k = 1, \ldots, n$. Then, $N = IJn$. Three cases can be considered.

1. H_0: $\delta_i = 0 (i = 1, \ldots, I)$ with no added assumptions about interactions. Then, $f_1 = I - 1$ and $f_2 = IJ(n-1)$. The numerator sum of squares is

   $$SSA = Jn\sum_i(\overline{Y}_{i..} - \overline{Y}_{...})^2 = \sum_i Y_{i..}^2/Jn - Y_{...}^2/(IJn)$$

   Since $E(\overline{Y}_{i..}) = \mu + \alpha_i$ and $E(\overline{Y}_{...}) = \mu$, it follows that $S^* = Jn\sum_i \delta_i^2$. Hence,

   $$\varphi = \sqrt{\frac{Jn\sum_i \delta_i^2}{I\sigma^2}} \tag{2.8}$$

2. H_0 is as in 1, but the interactions are assumed to be zero (i.e., $\rho_{ij} = 0$). Then, f_1, S^*, and φ are as in 1, but $f_2 = IJn - I - J + 1$

3. H_0: $\rho_{ij} = 0$ $(i = 1, \ldots, I;\ j = 1, \ldots, J)$. Then $f_1 = (I-1)(J-1)$ and $f_2 = IJ(n-1)$. The numerator sum of squares is

   $$SS(A \times B) = n\sum_i\sum_j(\overline{Y}_{ij.} - \overline{Y}_{i..} - \overline{Y}_{.j.} + \overline{Y}_{...})^2$$

   Since $E(\overline{Y}_{ij.}) = \mu + \delta_i + \rho_{ij}$ and $E(\overline{Y}_{.j.}) = \mu + \gamma_j$, it follows that $S^* = n\sum_i\sum_j \rho_{ij}^2$. Hence,

   $$\varphi = \sqrt{\frac{n\sum_i\sum_j \rho_{ij}^2}{[(I-1)(J-1)+1]\sigma^2}} \tag{2.9}$$

$$\text{Set } SSTr = \sum_i \sum_j (\overline{Y}_{ij.} - \overline{Y}_{...})^2 = \sum_i \sum_j Y_{ij.}^2/n - Y_{...}^2/(IJn)$$

$$\text{and } SST = \sum_i \sum_j \sum_k (Y_{ijk} - \overline{Y}_{...})^2 = \sum_i \sum_j \sum_k Y_{ijk}^2 - Y_{...}^2/(IJn)$$

The formula for SSB is obtained by interchanging i and j in SSA.

Table 2.9 is the $ANOVA$ table for cases 1 and 3.

Source	Sum of Squares	$d.f.$	$E(MS)$
Factor A	SSA	$I-1$	$\sigma^2 + Jn\sum_i \delta_i^2/(I-1)$
Factor B	SSB	$J-1$	$\sigma^2 + In\sum_i \gamma_i^2/(J-1)$
Interactions	$SS(A \times B) =$ $SSTr - SSA - SSB$	$(I-1)(J-1)$	$\sigma^2 + \frac{n\sum_i \sum_j \rho_{ij}^2}{(I-1)(J-1)}$
Treatments	$SSTr$	$IJ-1$	
Error	$SSE = SST - SSTr$	$IJ(n-1)$	σ^2
Total	SST	$IJn-1$	

Table 2.9

In the examples below each of the three cases will be considered.

The test statistic in case 1 is MSA/MSE and in case 3 the statistic is given by $MS(A \times B)/MSE$. In case 2 interactions and cells are deleted from the table, $SSE = SST - SSA - SSB$, and there are $IJn - I - J + 1$ degrees of freedom for error. The test statistic is MSA/MSE computed from the revised table.

EXAMPLE 2.3.1 (Effect of brain lesions on memory). Glick and Greenstein (1973) studied the effect of brain lesions on memory in mice. The mice were trained, by means of an electric shock, to avoid and withdraw from the entrance from one compartment to another. Immediately after training, electrodes were placed in the hippocampus or in the candate area of the brain. Electrical currents were then used to cause lesions. For some of the mice, however, no current was passed so that the surgical procedure was a sham. A second group of mice was treated in the same manner except that avoidance training did not involve shock; rather, it consisted of removal from the compartment.

The surgery was performed immediately after training. One hour after training the mice were placed in the apparatus again and the passage time from one compartment to the other was measured.

The model is given by (2.6) and (2.7). Let δ_i be the effect of the surgical procedure (hippocamus, candate, or sham) and let γ_j be the effect of the training method (shock or no shock). Then $I = 3$ and $J = 2$.

1. Set $\alpha = .05$ and let the power be .8 when $\sum_i \delta_i^2/\sigma^2 = 2$. Then $f_1 = 2$, $f_2 = 6(n-1)$, and (2.8) yields $\varphi = \sqrt{4n/3}$.

Table 2.10 displays the iteration. The critical value is $F_{.05}(2,18) = 3.555$

n	$\varphi = \sqrt{4n/3}$	$f_2 = 6(n-1)$	f_2 from chart
3	2.00	12	14
4	2.31	18	7

Table 2.10

2. Consider the same problem under the condition that the $\rho_{ij} = 0$. The iteration is still given in Table 2.10, the only modification being that $f_2 = 6n - 4$. Hence, $n = 3$ suffices. The critical value is $F_{.05}(2,14) = 3.739$

3. Set $\alpha = .05$ and let the desired power be .9 when $\sum_i \sum_j \rho_{ij}^2/\sigma^2 = 3$. Then, (2.9) yields $\varphi = \sqrt{n}$. Table 2.11 displays the iteration.

n	$\varphi = \sqrt{n}$	$f_2 = 6(n-1)$	f_2 from chart
4	2.00	18	∞
5	2.24	24	20

Table 2.11

The critical value for $n = 5$ is $F_{.05}(2,24) = 3.403$

In fact, Glick and Greenstein used varying numbers of mice (between ten and twelve) in each group.

EXAMPLE 2.3.2 (Reading Instruction). Burt and Lewis (1946) considered four different methods of reading instruction: alphabetic, kinesthetic, phonic, and visual. Pupils taught be each of the four methods were regrouped and given remedial reading by each of the four methods; that is, of pupils who originally received alphabetic instruction, four groups of n pupils each were formed, each group receiving one of the methods of remedial instruction. The same was repeated for pupils who originally received each of the other methods of instruction. The data may also be found in Walker and Lev (1953).

In this experiment, $I = J = 4$. The measurements are improvements in reading scores. For notation, we will assume that δ_i is the effect of the remedial instruction method, so that γ_j is the effect of the original instruction method.

Consider the three cases.

1. Let H_0 state that there is no main effect due to the method of remedial instruction. Then $f_1 = 3$ and $f_2 = 16(n-1)$. Let $\alpha = .005$ and the power be .6

when $\sum_i \delta_i^2/\sigma^2 = 4$. Then (2.8) yields $\varphi = 2\sqrt{n}$. The iteration is displayed in Table 2.12.

n	$\varphi = 2\sqrt{n}$	$f_2 = 16(n-1)$	f_2 from chart
1	2.00	0	25
2	2.83	16	7

Table 2.12

Burt and Lewis set $n = 3$. The critical value for $n = 2$ is $F_{.005}(3,16) = 6.303$

2. In this case $f_2 = 16n - 7$. With this change, the remainder of Table 2.12 is still appropriate and $n = 2$ suffices. The critical value is $F_{.005}(3,25) = 5.462$

3. Assume that $\alpha = .1$ and the power is to be .6 when $\sum_i \sum_j \rho_{ij}^2/\sigma^2 = 2$. Now $f_1 = 9, f_2 = 16(n-1)$, and (2.9) yields $\varphi = \sqrt{n/5}$. The iteration is displayed in Table 2.13.

n	$\varphi = \sqrt{n/5}$	$f_2 = 16(n-1)$	f_2 from chart
5	1.00	64	35

Table 2.13

The critical value is $F_{10}(9,64) = 1.731$

Alternative treatments of this problem are given in Secs. 2.8 and 2.9.

EXAMPLE 2.3.3 (Bioassay of vitamin C). As a means of comparing vitamin C intake, Crampton (1947) suggested measuring the lengths of odontoblasts in incisors of guinea pigs under three dosages of the unknown source, such as orange juice (also found in Bliss, 1952). Since sex seems to have a negligible effect, we will (as does Bliss) ignore the fact that half the animals in each of the six treatment groups would be males and half would be females. Letting δ_i be the effect of dosage form (control or unknown), we have $I = 2$, $J = 3$.

1. Set $\alpha = .01$ and let the power be .7 when $\sum_i \delta_i^2/\sigma^2 = 2$. Since $\delta_2 = -\delta_1$, this is equivalent to $\delta_i^2 = \sigma^2$. Now $f_1 = 1$, $f_2 = 6(n-1)$ and (2.8) yields $\varphi = \sqrt{3n}$. The iteration is displayed in Table 2.14.

n	$\varphi = \sqrt{3n}$	$f_2 = 6(n-1)$	f_2 from chart
2	2.45	6	13
3	3.00	12	7

Table 2.14

In fact, Crampton suggested $n \geq 10$. The critical value for $n = 3$ is $F_{.01}(1,12) = 9.330$

2. Consider the same problem with the interactions assumed to be zero. With this modification $f_2 = 6n - 4$, and n=3 still suffices. The critical value is $F_{.01}(1, 14) =$ 8.862

3. Assume that $\alpha = .001$ and set power equal to .9 when $\sum_i \sum_j \rho_{ij}^2/\sigma^2 = 3$. Now $f_1 = 2$, $f_2 = 6(n-1)$, and (2.9) yields $\varphi = \sqrt{n}$. The iteration is presented in Table 2.15 (Values of f_2 from the chart are very rough). The critical value is $F_{.001}(2, 54) = 7.872$ (See Chap. 6).

n	$\varphi = \sqrt{n}$	$f_2 = 6(n-1)$	f_2 from chart
7	2.65	36	200
8	2.83	42	100
9	3.00	48	60
10	3.16	54	40

Table 2.15

2.4 Randomized Blocks

Suppose that in replicating an experiment (taking more than one observation per cell) the additional observations fall in some natural way into blocks, one block containing a complete replication. In this case, the block itself may have some effect on the observation. This changes the resulting models.

For the one-way layout of Sec. 2.2, Eq. (2.2) is replaced by

$$Y_{ij} = \bar{\mu}_{..} + \delta_i + b_j + \epsilon_{ij} \tag{2.10}$$

Previous assumptions are augmented, without loss of generality, with $b_. = 0$. The b_j are the *block effects*. This is the model for the two-way layout with interactions absent and only one observation in each cell. Also, (2.3) is valid, $f_2 = (I-1)(n-1)$, and $N = In$. *SSTr* and *SST* are as in Sec. 2.2 so that S^* and φ are as there. Also, $SSB = \sum_j Y_{.j}^2/I - Y_{..}^2/(In)$.

Table 2.16 is the *ANOVA* table for this model. The test statistic is *MSTr/MSE*. Also, *SSE* is computed from *SST* − *SSTr* − *SSB*.

Source	Sum of Squares	*d.f.*	*E(MS)*
Treatments	*SSTr*	$I-1$	$\sigma^2 + n\sum_i \delta_i^2/(I-1)$
Blocks	*SSB*	$n-1$	$\sigma^2 + I\sum_j b_j^2/(n-1)$
Error	*SSE*	$(I-1)(n-1)$	σ^2
Total	*SST*	$In-1$	

Table 2.16

Similarly, for the two-way layout (2.6) is replaced by

$$Y_{ijk} = \bar{\mu}_{.} + \delta_i + \gamma_j + \rho_{ij} + b_k + \epsilon_{ijk} \tag{2.11}$$

Once again, previous assumptions are retained and in addition, we assume without loss of generality, that $b_. = 0$. This is the three-way layout with no interaction between blocks and other effects, and with one observation per cell. Also (2.8) is valid, $f_2 = (IJ - 1)(n - 1)$, and $N = IJn$.

The three cases of Sec. 2.3 are to be considered. The sums of squares are

$$SSA = \sum_i (\bar{Y}_{i..} - \bar{Y}_{...})^2 = \sum_i Y_{i..}^2/(Jn) - Y_{...}^2/(IJn)$$

with similar expressions for SSB and $SSBl$.

$$SSTr = \sum_i \sum_j (\bar{Y}_{ij.} - \bar{Y}_{...})^2 = \sum_i \sum_j Y_{ij.}^2/n - Y_{...}^2/(IJn)$$

$$SST = \sum_i \sum_j \sum_k (Y_{ijk} - \bar{Y}_{...})^2 = \sum_i \sum_j \sum_k Y_{ijk}^2 - Y_{...}^2/(IJn)$$

$$SS(A \times B) = SSTr - SSA - SSB$$

and

$$SSE = SST - SSTr - SSBl$$

For cases 1 and 3 the *ANOVA* table is shown in Table 2.17.

Source	SS	d.f.	E(MS)
Factor A	SSA	$I-1$	$\sigma^2 + Jn\sum_i \delta_i^2/(I-1)$
Factor B	SSB	$J-1$	$\sigma^2 + In\sum_j \gamma_j^2/(J-1)$
Interactions	$SS(A\times B)$	$(I-1)(J-1)$	$\sigma^2 + n\sum_i\sum_j \rho_{ij}^2/[(I-1)(J-1)]$
Treatments	$SSTr$	$IJ-1$	
Blocks	$SSBl$	$n-1$	$\sigma^2 + IJ\sum_k b_k^2/(n-1)$
Error	SSE	$(IJ-1)(n-1)$	σ^2
Total	SST	$IJn-1$	

Table 2.17

In case 1 the test statistic is MSA/MSE with $f_1 = I - 1$ while in case 3 it is $MS(A\times B)/MSE$ with $f_1 = (I-1)(J-1)$. In case 2 delete interactions and treatments, $SSE = SST - SSA - SSB - SSBl$, and $f_2 = (IJ-1)n - I - J + 2$. The test statistic is MSA/MSE, computed from the revised table. The expression for φ is unchanged.

EXAMPLE 2.4.1 (Word list learning). Kasschau (1972) studied the effect of meaningfulness on rates of learning of word lists. From a list of 96 words rated on a scale to reflect meaningfulness, twelve words each were selected from the low, middle, and high parts of this scale. The subjects observed each group of twelve words as often as needed to enable them to memorize seven words from the group. The number of trials required for each subject and group was observed. Here the subjects are the blocks and, in (2.17), the δ_i are the effects of meaningfulness.

Set $\alpha = .05$ and the power equal to .9 when $\sum_i \delta_i^2/\sigma^2 = 2$. Here $I = 3$ so that $f_1 = 2$, $f_2 = 2(n-1)$, and from (2.3), $\varphi = \sqrt{2n/3}$. The iteration is exhibited in Table 2.18.

n	$\varphi = \sqrt{2n/3}$	$f_2 = 2(n-1)$	f_2 from chart
6	2.00	10	∞
7	2.16	12	30
8	2.31	14	14

Table 2.18

In fact, Kasschau used 21 subjects. The critical value for $n = 8$ is $F_{.05}(2,14) = 3.739$

EXAMPLE 2.4.2 (Soybean yields). In a 1949 experiment at the Agronomy Farm, Ames, Iowa (see Ostle, 1963) each block contained early and late soybean plantings. Each of these plantings was divided into four portions and the portions were fertilized with one of Check, Aero, Na, or K. Let δ_i be the fertilizer effect.

For case 1, $f_1 = 3$ since $I = 4$. Furthermore, $J = 2$ so that $f_2 = 7(n-1)$. Let $\alpha = .01$ and the power be .8 when $\sum_i \delta_i^2 = \sigma^2/4$. Then, (2.8) yields $\varphi = \sqrt{n}/2$. The iteration is displayed in Table 2.19.

n	$\varphi = \sqrt{n}/2$	$f_2 = 7(n-1)$	f_2 from chart
16	2.00	105	200
17	2.06	112	80

Table 2.19

In the actual experiment, $n = 4$. For $n = 17$ the critical value $F_{.01}(3,112)$ is between $F_{.01}(3,120) = 3.949$ and $F_{.01}(3,90) = 4.007$. By interpolation (Chap. 6), $F_{.01}(3,112) = 3.961$

For case 2 we have $f_2 = 7n - 4$ so that the solution is unchanged.

For case 3, $f_1 = 3$. Suppose that $\alpha = .05$ and let the power be .7 when $\sum_i \sum_j \rho_{ij}^2 = 5\sigma^2/2$. The iteration is displayed in Table 2.20. By interpolation, $F_{.05}(3,63) = 2.751$

n	$\varphi = \sqrt{n}/2$	$f_2 = 7(n-1)$	f_2 from chart
9	1.50	56	200
10	1.58	63	30

Table 2.20

2.5 Latin Squares

Consider a three factor experiment, each factor being applied at m levels. The complete design requires m^3 observations per replicate. Assume that there is good reason to adopt the simplification of the general three-way layout model with no interactions. For a single replication this model is

$$(2.12) \qquad Y_{ijk} = \mu + \alpha_i + \beta_j + \gamma_k + \epsilon_{ijk} \quad (i, j, k = 1, \ldots, m)$$

where $\alpha_. = \beta_. = \gamma_. = 0$. When the model (2.12) is adopted it is possible to use a Latin square design which only requires m^2 observations per replicate.

An $m \times m$ *Latin square* is an $m \times m$ array consisting of m symbols arranged so that each symbol appears exactly m times, once in each row and once in each column. Table 2.21 illustrates a 4×4 Latin Square.

a	b	c	d
b	a	d	c
c	d	a	b
d	c	b	a

Table 2.21

In a Latin square design, the rows represent levels of factor A, the columns represent levels of factor B, and the small letters (symbols) represent levels of factor C. In (2.12) replace k by $k(i,j)$, the level of factor C when factors A and B are at levels i and j respectively. Thus, if the Latin square in Table 2.21 is used, $k(1,3) = 3$ while $k(2,3) = 4$.

Consider n replicates of an $m \times m$ Latin square so that $N = m^2 n$. Each replicate will use a different Latin square allowing an opportunity for a least a rough check of the assumption of no interactions. Let Y_{ijr} be the observation in the r-*th* replicate for levels i and j, respectively. Then

$$SSA = mn \sum_i (\overline{Y}_{i..} - \overline{Y}_{...})^2 = \sum_i Y_{i..}^2/(mn) - Y_{...}^2/(m^2 n)$$

with a similar expression for SSB.

Let $\tilde{Y}_k$ be the sum of all of the observations with factor C at level k and set $\overline{\tilde{Y}_k} = \tilde{Y}_k/(mn)$. Then,

$$SSC = mn\sum_k (\overline{\tilde{Y}_k} - \overline{Y}_{...})^2 = \sum_k \tilde{Y}_k^2/(mn) - Y_{...}^2/(m^2n)$$

Finally,

$$SST = \sum_i\sum_j\sum_r (Y_{ijr} - \overline{Y}_{...})^2 = \sum_i\sum_j\sum_r Y_{ijr}^2 - Y_{...}^2/(m^2n)$$

and

$$SSE = SST - SSA - SSB - SSC$$

Table 2.22 is the *ANOVA* table for this model.

Source	Sum of Squares	*d.f.*	*E(MS)*
Factor A	*SSA*	$m-1$	$\sigma^2 + mn\sum_i \alpha_i^2/(m-1)$
Factor B	*SSB*	$m-1$	$\sigma^2 + mn\sum_j \beta_j^2/(m-1)$
Factor C	*SSC*	$m-1$	$\sigma^2 + mn\sum_k \gamma_k^2/(m-1)$
Error	*SSE*	$m^2n - 3m + 2$	σ^2
Total	*SST*	$m^2n - 1$	

Table 2.22

If the null hypothesis is H_0: $\gamma_k = 0$ for $k = 1, \ldots, m$, then the test statistic is *MSC/MSE* with $f_1 = m - 1$ and $f_2 = m^2n - 3m + 2$. In this case

$$\varphi = \sqrt{\frac{n\sum_k \gamma_k^2}{\sigma^2}} \tag{2.13}$$

EXAMPLE 2.5.1 (Reactions to background music). Neter, Wasserman, and Kutner (1985) published data on a study to determine the effect of five types of background music on the productivity of tellers in a bank. The study was carried out daily for five one-week periods. A 5×5 Latin square design was used with the letters representing music types, rows representing weeks and columns representing days of the week. Suppose that H_0: $\gamma_k = 0$ $(k = 1,2,3,4,5)$ is to be tested with $\alpha = .01$ and it is desired that the power be .8 when $\sum_k \gamma_k^2/\sigma^2 = .5$. Then, (2.13) yields $\varphi = \sqrt{.5n}$. For $\alpha = .01$, power $= .8$, $f_1 = 4$, and $f_2 = \infty$, the value of $\varphi = 1.83$ so that initially $n = 7$. The iteration is displayed in Table 2.23.

n	$\varphi = \sqrt{.5n}$	$f_2 = 25n - 13$	f_2 from chart
7	1.87	162	200
8	2.00	187	40

Table 2.23

Thus, $n = 8$ suffices which requires a total of $N = 200$ observations.

A natural question is whether the total number of observations would be reduced if the complete model were replicated. The model is (2.12) with Y_{ijk} and ϵ_{ijk} replaced by Y_{ijkr} and ϵ_{ijk_r}, respectively, where $r = 1, \ldots, n$. Now

$$SSA = m^2 n \sum_i (\overline{Y}_{i\ldots} - \overline{Y}_{\ldots.})^2 = \sum_i Y_{i\ldots}^2/(m^2 n) - Y_{\ldots.}^2/(m^3 n)$$

with similar formulae for SSB and SSC. Also

$$SST = \sum_i \sum_j \sum_k \sum_r (Y_{ijkr} - \overline{Y}_{\ldots.})^2 = \sum_i \sum_j \sum_k \sum_r Y_{ijkr}^2 - Y_{\ldots.}^2/(m^3 n)$$

$$SSE = SST - SSA - SSB - SSC$$

and $N = m^3 n$. For this case, Table 2.24 is the *ANOVA* table.

Source	Sum of Squares	d.f.	E(MS)
Factor A	SSA	$m-1$	$\sigma^2 + m^2 n \sum_i \alpha_i^2/(m-1)$
Factor B	SSB	$m-1$	$\sigma^2 + m^2 n \sum_j \beta_j^2/(m-1)$
Factor C	SSC	$m-1$	$\sigma^2 + m^2 n \sum_k \gamma_k^2/(m-1)$
Error	SSE	$m^3 n - 3m + 2$	σ^2
Total	SST	$m^3 n - 1$	

Table 2.24

Then,

$$\varphi = \sqrt{\frac{mn \sum_k \gamma_k^2}{\sigma^2}} \tag{2.14}$$

In Example 2.5.1, $\varphi = \sqrt{2.5n}$ so that initially $n = 2$. The iteration is displayed in Table 2.25. By interpolation (Chap. 6), $F_{.01}(4, 187) = 3.421$ and $F_{.01}(4, 237) = 3.399$

n	$\varphi = \sqrt{2.5n}$	$f_2 = 125n - 13$	f_2 from chart
2	2.24	237	16

Table 2.25

With two replications of the complete design, 250 observations are required. Thus, eight replications of the Latin square design, requiring only 200 observations, is more economical.

Of course, if interactions among the factors cannot be ruled out, the analysis would be different. With all possible interactions in the model, for n replicates of the complete design, $f_2 = m^3(n-1)$ so that $m = 5$, $n = 2$ yields $f_2 = 125$, still adequate for the specifications of the example.

2.6 Simple Linear and Quadratic Regression

Suppose that there is a variable whose values can be set by the investigator (called the *independent variable*) and the effect of changes of this variable on some other variable (called the *dependent variable*) is to be examined. Assume that the effect is linear in the independent variable. However, the dependent variable cannot be observed directly, but is only observable with an added random fluctuation. The model is

$$Y_i = \delta_0 + \delta_1 x_i + \epsilon_i \quad (i = 1, \ldots, N) \tag{2.15}$$

where Y_i is the *i-th* observation of the dependent variable with random fluctuation added, x_i is the *i-th* value of the independent variable, the ϵ_i are independent and normally distributed with mean 0 and variance σ^2, and δ_0 and δ_1 are unknown constants.

The null hypothesis is H_0: $\delta_1 = 0$; that is, that Y_i does not depend on x_i or, in other words, that the Y_i are identically distributed. In this problem $f_1 = 1$ and $f_2 = N - 2$. From (2.15), $\mu_i = \delta_0 + \delta_1 x_i$ and if H_0 is true, then $\mu_i = \delta_0$.

Under the full model the estimators of δ_1 and δ_0 are respectively,
$\hat{\delta}_1 = (\sum_i x_i Y_i - x_. Y_./N)/(\sum_i x_i^2 - x_.^2/N)$ and $\hat{\delta}_0 = \overline{Y}_. - \hat{\delta}_1 \overline{x}_.$
Under H_0 the estimator of δ_0 is $\hat{\hat{\delta}}_0 = \overline{Y}_.$
Then, the numerator sum of squares is

$$SSH = \sum_i (\hat{\delta}_0 + \hat{\delta}_1 x_i - \hat{\hat{\delta}}_0)^2 = \hat{\delta}_1^2 (\sum_i x_i^2 - x_.^2/N)$$

Since $\hat{\delta}_1$ is an unbiased estimator, $S^* = \delta_1^2(\sum_i x_i^2 - x_.^2/N)$ and hence,

$$\varphi = \sqrt{\frac{\delta_1^2(\sum_i x_i^2 - x_.^2/N)}{2\sigma^2}} \tag{2.16}$$

Finally, $SST = \sum_i Y_i^2 - Y_.^2/N$, $SSE = SST - SSH$, and the test statistic is MSH/MSE. In the quadratic case, (2.15) is replaced by

$$Y_i = \delta_0 + \delta_1 x_i + \delta_2 x_i^2 + \epsilon_i \quad (i = 1, \ldots, N) \tag{2.17}$$

Assume that it is possible to code the x_i so that $x_. = \sum_i x_i^3 = 0$. This is easy to do if the x_i are equally spaced, since then they can be coded to be integers symmetric about 0.

Unbiased estimators of δ_2, δ_1, and δ_0 are respectively

$$\hat{\delta}_2 = (N\sum_i x_i^2 Y_i - Y.\sum_i x_i^2)/[N\sum_i x_i^4 - (\sum_i x_i^2)^2]$$

$$\hat{\delta}_1 = \sum_i x_i Y_i / \sum_i x_i^2, \text{ and } \hat{\delta}_0 = \overline{Y}. - \hat{\delta}_2 \sum_i x_i^2/n$$

Consider two cases:

1. H_0: $\delta_2 = 0$. Then, $f_1 = 1$ and $f_2 = N - 3$. Under H_0 the estimators $\hat{\hat{\delta}}_1$and $\hat{\hat{\delta}}_0$ of δ_0 and δ_1 respectively, are those for the full model in simple linear regression. Because of the coding, $\hat{\hat{\delta}}_1 = \hat{\delta}_1$ and $\hat{\hat{\delta}}_0 = \overline{Y}$.

The numerator sum of squares is

$$SSH = \sum_i(\hat{\delta}_0 + \hat{\delta}_1 x_i^2 + \hat{\delta}_2 - \hat{\hat{\delta}}_0 - \hat{\hat{\delta}}_1 x_i^2)^2 = \hat{\delta}_2^2[\sum_i x_i^4 - (\sum_i x_i^2)^2/N]$$

Hence, $S^* = \delta_2^2[\sum_i x_i^4 - (\sum_i x_i^2)^2/N]$ and

$$\varphi = \sqrt{\frac{\delta_2^2[\sum_i x_i^4 - (\sum_i x_i^2)^2/N]}{2\sigma^2}} \tag{2.18}$$

2. H_0: $\delta_1 = \delta_2 = 0$. Then, $f_1 = 2$ and $f_2 = N - 3$. The estimator of δ_0 under H_0 is $\hat{\hat{\delta}}_0 = \overline{Y}.$. The numerator sum of squares is

$$SSH = \sum_i(\hat{\delta}_0 + \hat{\delta}_1 x_i + \hat{\delta}_2 x_i^2 - \hat{\hat{\delta}}_0)^2 = \hat{\delta}_1^2 \sum_i x_i^2 + \hat{\delta}_2^2[\sum_i x_i^4 - (\sum_i x_i^2)^2/N]$$

so that $S^* = \delta_1^2 \sum_i x_i^2 + \delta_2^2[\sum_i x_i^4 - (\sum_i x_i^2)^2/N]$ and

$$\varphi = \sqrt{\frac{\delta_1^2 \sum_i x_i^2 + \delta_2^2[\sum_i x_i^4 - (\sum_i x_i^2)^2/N]}{3\sigma^2}} \tag{2.19}$$

EXAMPLE 2.6.1 (Differential pay rates). Valenzi and Andrews (1971) studied the effect of pay differentials on work output. A norm of \$1.40/hour was established as a pay rate for several workers doing the same job. Some were actually paid at this rate, while others were paid \$1.20/hour and still others \$2.00/hour. All workers knew the norm. Code the pay rates so that \$1.20 corresponds to $x_i = 0$, \$1.40 to $x_i = 1$, and \$2.00 to $x_i = 4$, that is $x_i = (R_i - 1.20)/.2$, where R_i is the pay rate for the *i-th* individual.

Assume the linear regression model (2.15), let $\alpha = .001$ and let the power be .995 when $\delta_1^2/\sigma^2 = 1$. If n subjects are tested at each pay level, then $N = 3n$ and, with the x_i coded as above, $x. = 5n$ and $\sum_i x_i^2 = 17n$ so that (2.16) yields $\varphi = \sqrt{13n/3}$. Furthermore, $f_2 = 3n - 2$. Table 2.26 displays the iteration.

n	$\varphi = \sqrt{13n/3}$	$f_2 = 3n - 2$	f_2 from chart
4	4.16	10	100
5	4.65	13	25
6	5.09	16	16

Table 2.26

In fact, Valenzi and Andrews used 31 subjects about evenly divided between the three pay scales. For $n = 6$ the critical value is $F_{.001}(1, 16) = 16.12$

Consider the quadratic model (2.17). In order to be able to code the x_i so that $x_. = \sum_i x_i^3 = 0$, suppose the three pay levels were \$1.20 (coded $x_i = -1$), \$1.40 ($x_i = 0$), and \$1.60 ($x_i = 1$). With n observations at each pay level $\sum_i x_i^2 = \sum_i x_i^4 = 2n$.

In case 1, set $\alpha = .001$, and the power equal to .995 when $\delta_2^2/\sigma^2 = 1$. Then (2.18) yields $\varphi = \sqrt{n/3}$. Table 2.27 summarizes the recursion.

n	$\varphi = \sqrt{n/3}$	$f_2 = 3(n-1)$	f_2 from chart
48	4.00	141	∞
49	4.04	144	300
50	4.08	147	200
51	4.12	150	150

Table 2.27

The critical value is $F_{.001}(1, 150) = 11.27$ by interpolation in Table 1.8 of Part Two.

In case 2, also set $\alpha = .001$, but let the power be .995 when $\delta_1^2 + \delta_2^2 = \sigma^2$. From (2.16) $\varphi = \sqrt{(2\delta_1^2 + 2\delta_2^2/3)n/3\sigma^2}$. The worst case is obtained by minimizing φ subject to $\delta_1^2 + \delta_2^2 = \sigma^2$. The minimum is obtained by setting $\delta_2^2 = \sigma^2$ and, then, $\varphi = \sqrt{2n}/3$. The recursion is summarized in Table 2.28.

n	$\varphi = \sqrt{2n}/3$	$f_2 = 3(n-1)$	f_2 from chart
54	3.58	159	100

Table 2.28

The critical value is $F_{.001}(2, 159) = 7.217$ by interpolation in Table 1.8 of Part Two.

In both of these cases interpolation in the chart is very rough and, for safety, one may wish to increase n slightly.

EXAMPLE 2.6.2 (Effect of vitamin E dosage on fertility). Mason (1942) administered several dosages of vitamin E on rats and observed the effect on fertility (data repeated in Bliss, 1952). The dosages administered were 3.75(1.25)7.5, 10.0, 15.0. The model assumed is (2.15) with the x_i representing the logarithms (base 10) of the dosages. If each dosage is applied to n rats, then $x_. = 5.12n$ and $\sum_i x_i^2 = 4.60n$. Since there are six different dosages $N = 6n$.

Test H_0: $\delta_1 = 0$ with $\alpha = .025$ and power equal to .95 when $\delta_1^2/\sigma^2 = 2$. Then (2.15) yields $\varphi = \sqrt{.231n}$. The recursion is summarized in Table 2.29.

n	$\varphi = \sqrt{.231n}$	$f_2 = 6n - 2$	f_2 from chart
28	2.54	166	100

Table 2.29

2.7 Multivariate t-tests

In this section all notation will be in terms of vectors and matrices.

Suppose that $\mathbf{X}_1, \mathbf{X}_2, \ldots, \mathbf{X}_N$ are observed p-dimensional random vectors which are independent and normally distributed with mean vector $\boldsymbol{\mu}$ and covariance matrix $\boldsymbol{\Sigma}$. The null hypothesis is H_0: $\boldsymbol{\mu} = \boldsymbol{\mu}_0$ and the alternative is H_1: $\boldsymbol{\mu} \neq \boldsymbol{\mu}_0$.

Let $T^2 = N(\overline{\mathbf{X}}_. - \boldsymbol{\mu}_0)'\mathbf{S}^{-1}(\overline{\mathbf{X}}_. - \boldsymbol{\mu}_0)$ where $\mathbf{S} = \sum_i(\mathbf{X}_i - \overline{\mathbf{X}}_.)(\mathbf{X}_i - \overline{\mathbf{X}}_.)'/(N-1)$. The analogue of the t-test, proposed by Hotelling (1931), rejects when T^2 is large. Hotelling proved that the distribution of $T^2(N - p + 1)/[(N-1)p]$ is noncentral F with $f_1 = p$, $f_2 = N - p$, and

$$\varphi = \sqrt{N(\boldsymbol{\mu} - \boldsymbol{\mu}_0)'\,\boldsymbol{\Sigma}^{-1}(\boldsymbol{\mu} - \boldsymbol{\mu}_0)/(p+1)} \tag{2.20}$$

Under H_0, the distribution is central F since $\varphi = 0$.

EXAMPLE 2.7.1 (Metabolism of flouroacetate). Preuss et. al. (1968) studied the absorption of fluoracetate by Acacia georginae. They provided the plants with radioactive fluoracetate and measured the radioactivity in five lipid fractions. Thus, for each plant they observed a five-dimensional vector. This is, of course, also the dimensionality of $\boldsymbol{\mu}$. Suppose that, after coding the data, the problem is to test H_0: $\boldsymbol{\mu} = \mathbf{0}$ against H_1: $\boldsymbol{\mu} \neq \mathbf{0}$.

Set $\alpha = .1$ and from the power equal to .7 if $\boldsymbol{\mu}'\,\boldsymbol{\Sigma}^{-1}\boldsymbol{\mu} = 1$. In this case, $f_1 = 5$, $f_2 = N - 5$, and from (2.20), $\varphi = \sqrt{N/6}$. Table 2.30 displays the recursion.

N	$\varphi = \sqrt{N/6}$	$f_2 = N-5$	f_2 from chart
9	1.22	4	60
10	1.29	5	25
11	1.35	6	17
12	1.41	7	11
13	1.47	8	10
14	1.53	9	8

Table 2.30

The critical value is $F_{.10}(5,9) = 2.611$

See Secs. 2.8 and 2.9 for alternative treatments of this problem.

The test given above is the multivariate extension of the usual one-sample t-test. A similar extension of the two-sample t-test is possible. The observable p-dimensional vectors are $\mathbf{X}_1^{(1)},\ldots,\mathbf{X}_{N_1}^{(1)}$ and $\mathbf{X}_1^{(2)},\ldots,\mathbf{X}_{N_2}^{(2)}$ which are independent, normally distributed, all with covariance matrix $\boldsymbol{\Sigma}$. For $i = 1,\ldots,N_j$, the $\mathbf{X}_i^{(j)}$ all have mean $\boldsymbol{\mu}^{(j)}$ $(j = 1,2)$. The null hypothesis is H_0: $\boldsymbol{\mu}^{(1)} = \boldsymbol{\mu}^{(2)}$ and the alternative is H_1: $\boldsymbol{\mu}^{(1)} \neq \boldsymbol{\mu}^{(2)}$. Set

$$\overline{\mathbf{X}}_{\cdot}^{(j)} = \sum_i \mathbf{X}_i^{(j)}/N_j \ (j = 1,2)$$

$$\mathbf{S} = [\sum_{i=1}^{N_1}(\mathbf{X}_i^{(1)} - \overline{\mathbf{X}}_{\cdot}^{(1)})(\mathbf{X}_i^{(1)} - \overline{\mathbf{X}}_{\cdot}^{(1)})' + \sum_{i=1}^{N_2}(\mathbf{X}_i^{(2)} - \overline{\mathbf{X}}_{\cdot}^{(2)})(\mathbf{X}_i^{(2)} - \overline{\mathbf{X}}_{\cdot}^{(2)})']/(N_1 + N_2 - 2)$$

and

$$T^2 = \frac{N_1 N_2}{N_1 + N_2}(\mathbf{X}_{\cdot}^{(1)} - \mathbf{X}_{\cdot}^{(2)})'\mathbf{S}^{-1}(\mathbf{X}_{\cdot}^{(1)} - \mathbf{X}_{\cdot}^{(2)})$$

The distribution of $T^2(N_1+N_2-p-1)/[(N_1+N_2-2)p]$ is noncentral F with $f_1 = p$, $f_2 = N_1 + N_2 - p - 1$, and

$$\varphi = \sqrt{N_1N_2(\boldsymbol{\mu}^{(1)} - \boldsymbol{\mu}^{(2)})'\boldsymbol{\Sigma}^{-1}(\boldsymbol{\mu}^{(1)} - \boldsymbol{\mu}^{(2)})/[(N_1 + N_2)(p+1)]} \tag{2.21}$$

Under H_0, $\varphi = 0$ so that the distribution is central F.

EXAMPLE 2.7.2 (Iris blossom dimensions). Fisher (1936) presented four measurements (sepal length, sepal width, petal length, and petal width) on each member of a sample of blossoms of Iris versicolor (first sample) and of Iris setosa (second sample). Let $\mathbf{X}_i^{(j)}$ be the vector of observations (in the order given above for the i-th blossom ($i = 1,\ldots,n$) from the j-th sample ($j = 1,2$).

Test H_0 with $\alpha = .1$ and power .995 when $(\boldsymbol{\mu}^{(1)} - \boldsymbol{\mu}^{(2)})'\boldsymbol{\Sigma}^{-1}(\boldsymbol{\mu}^{(1)} - \boldsymbol{\mu}^{(2)}) = .1$ Set $N_1 = N_2 = n$ so that $f_1 = 4$, $f_2 = 2n - 5$, and (2.21) yields $\varphi = \sqrt{n/10}$. The recursion is exhibited in Table 2.31.

n	$\varphi = \sqrt{n/10}$	$f_2 = 2n - 5$	f_2 from chart
49	2.21	93	∞
50	2.24	95	200
51	2.26	97	150
52	2.28	99	80

Table 2.31

In fact, in Fisher's data $n = 50$. The critical value for $n = 52$ is $F_{.10}(4, 99) = 2.003$

Now consider the profile analysis problem. The setup is that of the two-sample multivariate t-test just considered. Let $\mu_i^{(j)}$ be the i-*th* coordinate of $\mu^{(j)}$. The problem is to test

$$H_0 \colon \mu_i^{(1)} - \mu_{i+1}^{(1)} = \mu_i^{(2)} - \mu_{i+1}^{(2)} \ (i = 1, \ldots, p-1)$$

against the general alternative. Let

$$T^2 = \frac{N_1 N_2}{N_1 + N_2} (\overline{\mathbf{X}}^{(1)} - \overline{\mathbf{X}}^{(2)})' \mathbf{C}' (\mathbf{C}\mathbf{S}\mathbf{C}')^{-1} \mathbf{C} (\overline{\mathbf{X}}^{(1)} - \overline{\mathbf{X}}^{(2)})$$

where $\mathbf{C}$ is the $(p-1) \times p$ matrix

$$\mathbf{C} = \begin{bmatrix} 1 & -1 & 0 & 0 & 0 & \ldots & 0 & 0 & 0 \\ 0 & 1 & -1 & 0 & 0 & \ldots & 0 & 0 & 0 \\ \vdots & \vdots & \vdots & \vdots & \vdots & & \vdots & \vdots & \vdots \\ 0 & 0 & 0 & 0 & 0 & \ldots & 0 & 1 & -1 \end{bmatrix}$$

Note that $\mathbf{C}(\mu^{(1)} - \mu^{(2)})$ is the $(p-1)$-dimensional column vector with i-*th* component $(\mu_i^{(1)} - \mu_{i+1}^{(1)}) - (\mu_i^{(2)} - \mu_{i+1}^{(2)})$. Then, $T^2(N_1 + N_2 - p - 1)/[(N_1 + N_2 - 2)p]$ has the noncentral F-distribution with $f_1 = p - 1$, $f_2 = N_1 + N_2 - p$, and

$$\varphi = \sqrt{\frac{N_1 N_2}{p(N_1 + N_2)} (\mu^{(1)} - \mu^{(2)})' \mathbf{C}' (\mathbf{C}\mathbf{S}\mathbf{C}')^{-1} \mathbf{C} (\mu^{(1)} - \mu^{(2)})} \tag{2.22}$$

Under H_0: $\varphi = 0$.

EXAMPLE 2.7.3. For the measurements of Example 2.7.2, consider the profile analysis problem and set $N_1 = N = n$. Then, $f_1 = 3$, $f_2 = 2n - 4$ and

$$\mathbf{C} = \begin{bmatrix} 1 & -1 & 0 & 0 \\ 0 & 1 & -1 & 0 \\ 0 & 0 & 1 & -1 \end{bmatrix}$$

Let $\alpha = .05$ and the power be .9 when

$$(\mu^{(1)} - \mu^{(2)})' \mathbf{C}' (\mathbf{C} \, \mathbf{\Sigma} \mathbf{C}')^{-1} \mathbf{C} (\mu^{(1)} - \mu^{(2)}) = .5$$

Then (2.22) yields $\varphi = \sqrt{n}/4$. Table 2.32 summarizes the iteration.

n	$\varphi = \sqrt{n}/4$	$f_2 = 2n - 4$	f_2 from chart
58	1.90	112	150
59	1.92	114	70

Table 2.32

2.8 Fixed α on a Sphere

In all the examples considered so far a null hypothesis, denoted by H_0, that certain parameters are zero is tested against the general alternative, H_1. In fact, the real interest is usually in testing the hypothesis that the parameters are small against the alternative that they are not. From the form of S^* in the examples, the power is constant on spheres of dimension equal to the numbers of parameters involved in the hypothesis. The radius of the sphere is always a multiple of the unknown parameter σ.

Suppose that the hypothesis of interest H', that this radius is at most $\Delta'\sigma$, is to be tested with level of significance α'. For any sample size this is equivalent to testing H_0 at some level of significance $\alpha < \alpha'$ using a test with power at most α' on the sphere of radius $\Delta'\sigma$. Suppose that the power on a sphere of radius $\Delta\sigma$ ($\Delta \geq \Delta'$) is fixed. Proceed as follows:

1. For each $\alpha < \alpha'$ find n to guarantee the desired power on the sphere of radius $\Delta\sigma$.

2. If n yields power less than α' on the sphere of radius $\Delta'\sigma$, then the desired properties are achieved by testing H_0 at level of significance α using that n. Otherwise, that α may not be used.

3. Of those values of α passing the test in (2), choose the one that minimizes n.

As α is decreased, n will increase. Hence, the α's are chosen in decreasing order until a value is found which passes the test in (2). Note that the solution obtained will only be approximate. It may happen that the correct value of α is not among those for which charts exist. A smaller value of n may be obtainable through computation of additional charts.

The test in (2) requires the comparison of the computed f_2 with f_2 obtained for power α' obtained from the charts. If the computed f_2 is smaller, then α and n pass the test.

Several of the previous examples will be reexamined from this point of view. The numbering in parentheses is that of the original examples. The notation sets $\varphi' = \sqrt{S^*/[(f_1+1)\sigma^2]}$ with S^* evaluated on the sphere of radius $\Delta'\sigma$.

EXAMPLE 2.8.1 (Example 2.2.1). Now test H': $\delta_1^2 \le \sigma^2/5$ against H_1: $\delta_1^2 > \sigma^2/5$ with $\alpha' = .05$. As before, let the power be .8 when $\delta_1^2 = 2\sigma^2$ so that (2.3) yields $\varphi = \sqrt{2n}$. Furthermore, $\varphi' = \sqrt{n/5}$.

In this example and in the remaining examples of this section the solutions for n will be omitted. These follow the previous pattern. Table 2.33 summarizes the remainder of the solution.

α	n	$f_2 = 2(n-1)$	$\varphi' = \sqrt{n/5}$	Required f_2
.025	4	6	0.89	<4
.010	5	8	1.00	<4
.005	6	10	1.10	7
.001	8	14	1.26	50

Table 2.33

Hence, the solution is $\alpha = .001$, $n = 8$ and the critical value is $F_{.001}(1, 14) = 17.14$

EXAMPLE 2.8.2 (Example 2.3.2). Return to case 1 and assume the same power requirement so that $\varphi = 2\sqrt{n}$. Let H': $\sum_i \delta_i^2/\sigma^2 \le 1/4$ be tested with $\alpha = .005$. Thus, $\varphi' = \sqrt{n/2}$ from (2.8). Only $\alpha = .001$ is available. This results in $n = 2$ so that $f_2 = 16$. Thus, $\varphi' = 1$ which requires $f_2 = 4$ for power at most .005 when $\sum_i \delta_i^2/\sigma^2 = 1/4$. This does not suffice, and for a complete solution of the problem, charts would be needed for $\alpha < .001$.

EXAMPLE 2.8.3 (Example 2.6.2). Consider the model in (2.15) and the problem of testing H': $\delta_1^2/\sigma^2 \le .1$ with $\alpha' = .025$ and power equal to .9 when $\delta_1^2/\sigma^2 = 2$. Then, $\varphi = \sqrt{.231n}$ and $\varphi' = \sqrt{.0115n}$. The results are summarized in Table 2.34.

α	n	$f_2 = 6n - 2$	$\varphi' = \sqrt{.0115n}$	Required f_2
.010	33	196	.62	4
.005	34	202	.63	18
.001	43	256	.70	∞

Table 2.34

The solution is $\alpha = .001$, $n = 43$ and the critical value is $F_{.001}(1, 256) = 11.08$

EXAMPLE 2.8.4 (Example 2.7.1). Consider H': $\mu' \Sigma^{-1} \mu \le .01$. Set $\alpha' = .1$ and the power equal to .7 if $\mu' \Sigma^{-1} \mu = 1$. In this example the power is constant on ellipsoids which reduce to spheres if $\Sigma = \sigma^2\mathbf{I}$ where $\mathbf{I}$ is the $p \times p$ identity matrix. Then, $\varphi = \sqrt{N/6}$ and $\varphi' = \sqrt{N/600}$. For $\alpha = .05$, the requirement is that $N = 17$ so that $\varphi' = .17$. For power equal to .1 this requires that $f_2 = \infty$ so that $\alpha = .05$ with $N = 17$ suffices. The critical value is $F_{.05}(5, 12) = 3.106$

2.9 A Bayesian Approach

Suppose that prior probabilities are assigned: ξ to H_0 and $1 - \xi$ to H_1. Assume, further, that each observation costs C, the loss due to rejecting a true hypothesis is A, and the loss due to accepting false hypothesis is B. The Bayes risk of a test is

$$\rho = CN + A\alpha\xi + B(1 - \text{power})(1 - \xi) \tag{2.23}$$

An F-test Bayes in the class of F-tests may be found by obtaining N for each (α, power) combination and choosing the combination which minimizes ρ. Of course, with any set of charts, this can only be approximated.

There is an additional slight source of error here. The risks obtained through the charts are, in fact, somewhat larger than the actual values. The value of N found for a given α and power actually yields a true power somewhat larger. This can be seen from the first example below. For $\alpha = .001$, we require $n = 3$ for power .005. But, $n = 3$ also guarantees power .01.

Is this test Bayes in the class of all tests? The answer is yes provided we assign a uniform prior distribution on the sphere and we assign to σ the improper density $\lambda(\sigma) = \sigma^{-1}$ if $\sigma > 0$, 0 if $\sigma \leq 0$. The fact that the F-test is Bayes then follows from the fact that it is best invariant under the appropriate group of transformations.

The methods of this section may be generalized to the case in which more than one sphere carries positive prior probability.

We will illustrate this approach using the same examples used in Sec. 2.8 and number those examples according to their original appearances in previous sections.

EXAMPLE 2.9.1 (Example 2.2.1). Assume that the cost of each animal tested is $C = 100$ and the costs of errors are $A = B = 10{,}000$. Assume prior probabilities 1/2 each for H_0 and H_1: $\delta_1^2 = 2\sigma^2$. Recalling that $N = 2n$, (2.23) yields $\rho = 200n + 5000(\alpha + 1 - \text{power})$. Table 2.35 summarizes the procedure. Any no observation decision has α =power and hence, $\rho = 5000$.

The asterisk indicates the smallest risk for that α. Examination of Table 2.35 shows that the smallest risk, 1450, is obtained by setting $\alpha = .025$, and power equal to .975, which requires $n = 6$. The critical value is $F_{.025}(1, 10) = 6.937$

For $\alpha = .001$ no power less than .1 need be considered since power .05 and $n = 2$ yield $\rho = 5155$ which is greater than the no observation risk. Similarly, $n > 10$ need never be considered since, with $n = 11$, the first term in ρ is 2220, which is greater than the risk for $\alpha = .001$, power .95. As smaller risks are found, even more restrictions may be placed on the range of power and n examined. Finally, $\alpha \geq .25$ need not be considered since, in that case, $\rho > 5000\alpha + 400 \geq 1650$.

$\alpha = .001$			$\alpha = .005$			$\alpha = .01$		
Power	n	ρ	Power	n	ρ	Power	n	ρ
.100	4	5305	.700	5	2525	.800	5	2050
.200	5	5005	.800	6	2225	.900	6	1750
.300	5	4505	.900	7	1925	.950	7	1700*
.400	5	4005	.950	8	1875	.975	8	1775
.500	5	3705*	.975	8	1750*	.990	8	1700*
$\alpha = .025$			$\alpha = .050$			$\alpha = .10$		
Power	n	ρ	Power	n	ρ	Power	n	ρ
.800	4	1925	.900	4	1550	.900	4	1800
.900	5	1625	.950	5	1500*	.950	4	1550*
.950	6	1575	.975	6	1575	.975	5	1625
.975	6	1450*	.990	6	1500*	.990	5	1550*

Table 2.35

For the examples that follow the procedure will be summarized by only tabling α, the power yielding the smallest risk, the corresponding n, and the risk.

EXAMPLE 2.9.2 (Example 2.3.2). Assume in case 1 that $C = 1, A = 500, B = 200$, and the prior probabilities are 1/3 for H_0 and 2/3 for H_1: $\sum_i \delta_i^2/\sigma^2 = 4$. Since $N = 16n$, from (2.23) $\rho = 16n + [500\alpha + 400(1 - \text{power})]/3$. Table 2.36 summarizes the method.

α	Optimal power	n	ρ
.001	.975	3	51.5
.005	.995	3	49.5
.010	.900	2	47.0
.025	.975	2	39.5

Table 2.36

Thus, set $n = 2$, $\alpha = .025$, and the power will be .975. The risk is then $\rho = 39.5$. The critical value is $F_{.025}(3, 16) = 4.077$

EXAMPLE 2.9.3 (Example 2.6.2). Each observation requires an expensive rat so let $C = 10$. It seems reasonable that the dosage has an effect on the fertility, so set prior probabilities 1/4 for H_0 and 3/4 for the alternative. This being a purely scientific study, the two errors are equally important, casting doubt on the investigators qualifications. Hence, set $A = B = 2000$. Since $N = 6n$, (2.23) yields

$$\rho = 60n + 500[\alpha + 3(1 - \text{power})]$$

The optimal decision with $n = 0$ is to always reject H_0. In that case, $\alpha =$ power $= 1$ so that $\rho = 500$. Continuing as usual yields Table 2.37.

Since the charts for $\alpha = .5$ are not usable for $f_1 = 1$, continuation is impossible. Of the cases considered the rule for $n = 0$ (always reject) has the smallest risk.

α	Optimal power	n	ρ
.001	–	≥32	>1920
.005	–	≥24	>1440
.010	–	≥21	≥1260
.025	–	≥16	>960
.050	–	≥13	>780
.100	–	≥10	>600
.250	.8	9	965

Table 2.37

EXAMPLE 2.9.4 (Example 2.7.1). Assume that the cost per observation (per Acacia tree) is $C = 1$ and, as an expression of confidence that the hypothesis is false, assign prior probabilities 1/10 to H_0 and 9/10 to H_1: $\mu' \Sigma^{-1} \mu = 1$. Suppose that the errors are of equal importance with $A = B = 10000$. Then (2.23) yields $\rho = N + 100[\alpha + 9(1 - \text{power})]$. Following the usual procedure yields Table 2.38.

α	Optimal power	n	ρ
.001	.995	57	61.6
.005	.995	50	55.0
.010	.995	45	50.5
.025	.995	40	47.0
.050	.995	35	44.5
.100	.995	32	46.5
.250	.995	25	54.5

Table 2.38

The optimal procedure sets $\alpha = .05$ with $N = 35$ to achieve power .995. This procedure has risk $\rho = 44.5$. The critical value is $F_{.05}(5, 30) = 2.534$

Chapter 3

BACKGROUND

This chapter begins with a general discussion of hypothesis testing and follows that with the specialization to general linear models.

3.1 Tests of Hypotheses

The following formulation of the hypothesis testing problem is due to Neyman and Pearson (1928).

Let $\boldsymbol{Y}$ be the observable random data. If the observable data consist of N numerical observations, then the representation is in the form $\boldsymbol{Y} = (Y_1, Y_2, \ldots, Y_N)$, that is, as a random point in N-dimensional Euclidean space.

Suppose that the distribution of $\boldsymbol{Y}$ depends on the *unknown parameter* $\boldsymbol{\theta} = (\theta_1, \theta_2, \ldots, \theta_r)$. A problem of choosing an action based on $\boldsymbol{Y}$ with consequences depending on $\boldsymbol{\theta}$ is a *statistical problem*. Denote the set of possible values of $\boldsymbol{\theta}$ by $\boldsymbol{\Theta}$.

In this setup, $Y_1, Y_2, \ldots, Y_N$ may be independent random variables, normally distributed with mean μ and variance σ^2. Suppose that μ and σ^2 are unknown so that $\boldsymbol{\theta} = (\mu, \sigma^2)$, or equivalently $(\boldsymbol{\theta} = (\mu, \sigma))$. Any problem of choosing an action based on $\boldsymbol{Y}$ with consequences depending on μ and σ^2 would be a statistical problem. Also, $\boldsymbol{\Theta}$ is a point in the half plane $\sigma^2 > 0$ (or $\sigma > 0$).

A *hypothesis* is a statement that $\boldsymbol{\theta}$ lies in some well-defined subset $\boldsymbol{\Theta}_0$ of $\boldsymbol{\Theta}$. A *hypothesis testing problem* is a statistical problem in which one is to decide if the hypothesis that $\boldsymbol{\theta}$ is in $\boldsymbol{\Theta}_0$ (called the *null hypothesis*) is true. The problem is often stated in the form H_0: $\boldsymbol{\theta} \in \boldsymbol{\Theta}_0$ vs H_1: $\boldsymbol{\theta} \in \boldsymbol{\Theta}_1$ where $\boldsymbol{\Theta}_1$ is the complement of $\boldsymbol{\Theta}_0$ with respect to $\boldsymbol{\Theta}$. Then, H_1 is called the *alternative hypothesis*. Another statement of the problem is H_0: $\boldsymbol{\theta} \in \boldsymbol{\Theta}_0$ vs H_1: $\boldsymbol{\theta} \notin \boldsymbol{\Theta}_0$.

In the example, let $\Theta_0 = \{(5, \sigma^2) : \sigma^2 > 0\}$. This problem would be stated as H_0: $\mu = 5$ vs H_1: $\mu \neq 5$.

In all hypothesis testing problems there are two possible actions, *accept* H_0 (behave as if H_0 is true) and *reject* H_0 (behave as if H_1 is true).

A *test* is a rule that specifies for each $\boldsymbol{Y}$ whether H_0 will be accepted or rejected. In the example, the usual t-test examines the quantity

$$t = \frac{\overline{Y} - 5}{s/\sqrt{N}}$$

where $\overline{Y} = \sum_{i=1}^{N} Y_i/N$ and $s^2 = \sum_{i=1}^{N}(Y_i - \overline{Y})^2/(N-1)$. If $|t|$ is sufficiently large, this test rejects H_0. Otherwise it accepts H_0. Note that t is computable from $\boldsymbol{Y} = (Y_1, Y_2, \ldots, Y_N)$ so that the t-test is a rule of the type described.

Errors occur in two cases: (1) when $\boldsymbol{\theta} \in \Theta_0$ and H_0 is rejected and (2) when $\boldsymbol{\theta} \in \Theta_1$ and H_0 is accepted. This leads to the definition of two types of errors as given in Table 3.1.

	State of Nature	
Action	$\boldsymbol{\theta} \in \Theta_0$	$\boldsymbol{\theta} \in \Theta_1$
Accept H_0	Correct action taken	Type II error committed probability = operating characteristic
Reject H_0	Type I error committed probability = α	Correct action taken probability = power

Table 3.1

In the example, a type I errors occurs if H_0 is rejected when, in fact, the mean is 5, while the error of accepting H_0 if the mean is not 5 is a type II error.

It is desirable to control the probabilities of both kind of errors. In the fixed sample size case, a decrease in one of the error probabilities results in an increase in the other. Thus, a balance must be struck. The typical procedure is to fix a small upper bound α on the probability of a type I error and then find a test which minimizes the probability of a type II error.

The tests considered here have the property that the probability of a type I error does not depend on the particular value that $\boldsymbol{\theta}$ takes in Θ_0. For example, for the t-test the probability of a type I error does not depend on σ^2. Thus, in fact, $\alpha = P$(type I error) in these cases and is called the *level of significance.*

Equivalent to minimizing the probability of a type II error is maximizing the probability of rejecting the null hypothesis when $\boldsymbol{\theta}$ is in $\boldsymbol{\Theta}_1$. The latter probability is called the *power* of the test. The power does usually depend on the the value that $\boldsymbol{\theta}$ takes in $\boldsymbol{\Theta}_1$. The power of the t-test depends on $(\mu - 5)^2/\sigma^2$.

The power of a test also will depend on the sample size. Intuitively one expects, as is the case, that power increases as the sample size increases. It is the purpose of the charts presented in Part Three to make possible the choice of the sample size when α and the power are preset for problems in the class of general linear hypotheses.

3.2 The General Linear Hypothesis

We now give a statement of the problem of general linear hypothesis.

Assume that $\boldsymbol{Y} = (Y_1, Y_2, \ldots, Y_N)$ is observed where

$$Y_i = \beta_0 + \beta_1 x_{1i} + \cdots + \beta_k x_{ki} + \epsilon_i \tag{3.1}$$

and $\beta_0, \beta_1, \ldots, \beta_k$ are unknown, fixed parameters; the x_{ji} $(i = 1, \ldots, N;\ j = 1, \ldots, k)$ are constants determined by the condition of the *i-th* trial (and, hence, known and fixed by the experimenter); and the ϵ_i are independent, normally distributed random variables with mean 0 and variance σ^2.

It should be noted that the model is linear in the β_j since one possibility is that $x_{ji} = x_{1i}^j$. As an example, consider the effect of room temperature on the life of a light bulb. Letting Y_i be the life of the *i-th* bulb and x_{1i} be the room temperature at which it is tested (at the control of the experimenter), then the form of (3.1) in this case might be

$$Y_i = \beta_0 + \beta_1 x_{1i} + \beta_2 x_{1i}^2 + \beta_3 x_{1i}^3 + \epsilon_i \tag{3.2}$$

cubic regression on temperature.

The hypothesis to be tested asserts that certain linear linear relationships hold among the β_j's. For example, in (3.2) the hypothesis to be tested might be that $\beta_1 = \beta_2 = \beta_3 = 0$, (requiring three equations), that is, that room temperature has no effect on the life of a light bulb. Another possibility is H_0: $\beta_1 + 2\beta_2 + 3\beta_3 = 0$.

The traditional test used is the F-test. Its power depends on N and on the noncentrality parameter, to be discussed. The charts enable one to choose N to attain a desired power under fixed conditions. A variety of examples of F-tests were given in Chap. 2.

An assumption is made that the $\boldsymbol{x}_j = (x_{j1}, \cdots, x_{jN})$ are linearly independent vectors, that is, that there do not exist constants $c_1, \ldots, c_k$ which are not all zero such that $c_1 x_{1i} + \ldots + c_k x_{ki} = 0$ for all $i = 1, \ldots, N$. If this assumption were not true and, for example $c_k \neq 0$, then $x_{ki} = -(c_1 x_{1i} + \cdots + c_{k-1} x_{k-1})/c_k$ and the x_{ki} can be removed from the model. This can continue until the linear independence assumption is satisfied, the model being reduced at each step.

Let $\mu_i = E(Y_i) = \beta_0 + \beta_1 x_{1i} + \cdots + \beta_k x_{ki}$ and $\boldsymbol{\mu} = (\mu_1, \ldots, \mu_N)$. Denote the N-dimensional Euclidean space by $\mathcal{R}^N$ and let $\boldsymbol{\Theta}^{(\boldsymbol{\mu})}$ be the set of $\boldsymbol{\mu}$ given by (3.1). Then, $\boldsymbol{\Theta} = \boldsymbol{\Theta}^{(\boldsymbol{\mu})} \times (0, \infty)$ where $\boldsymbol{\Theta}^{(\boldsymbol{\mu})}$ is a $(k+1)$-dimensional subspace of $\mathcal{R}^N$ (a hyperplane containing the origin). Here $(0, \infty)$ stands for $\{\sigma^2 : \sigma^2 > 0\}$ and represents the fact that any $\sigma^2 > 0$ (or $\sigma > 0$) is possible. The symbol $\times$ stands for "Cartesian product."

The hypothesis H_0: $\beta_{l+1} = \beta_k$ can be expressed as H_0: $\boldsymbol{\mu} \in \boldsymbol{\Theta}_0^{(\boldsymbol{\mu})}$ where $\boldsymbol{\Theta}_0^{(\boldsymbol{\mu})}$ is an $(l+1)$-dimensional subspace which is contained in $\boldsymbol{\Theta}^{(\boldsymbol{\mu})}$.

3.3 Least Squares Estimates and the F-test

For any $\mathbf{z} = (z_1, z_2, \ldots, z_N)$, let $||\mathbf{z}||^2 = \sum_{i=1}^N z_i^2$. Consider $S = ||\boldsymbol{Y} - \boldsymbol{\mu}||^2$ and let $\hat{\boldsymbol{\mu}}$ be the point in $\boldsymbol{\Theta}^{(\boldsymbol{\mu})}$ which minimizes S. Then $\hat{\boldsymbol{\mu}}$ is the *least squares estimate* (*LSE*) of $\boldsymbol{\mu}$ and is an unbiased estimate ($E(\hat{\boldsymbol{\mu}}) = \boldsymbol{\mu}$). Similarly, let $\hat{\hat{\boldsymbol{\mu}}}$ be the *LSE* of $\boldsymbol{\mu}$ assuming that H_0 is true.

Set $SSE = ||\boldsymbol{Y} - \hat{\boldsymbol{\mu}}||^2$ and $SSH = ||\hat{\boldsymbol{\mu}} - \hat{\hat{\boldsymbol{\mu}}}||^2$, called the *sum of squares for error* and the *sum of squares for the hypothesis* H_0, respectively. Then, $f_2 = N - k - 1$ is the *number of degrees of freedom for error* and $MSE = SSE/f_2$ is the *mean square for error*. Note that MSE measures how closely the model in (3.1) fits the data. Let $f_1 = k - l$, the *number of degrees of freedom for the hypothesis* H_0 and $MSH = SSH/f_1$, the *mean square for the hypothesis* H_0. Then, MSH measures the loss of fit of $\boldsymbol{\mu}$ to the data if H_0 is assumed to be true.

Let $\mathcal{F} = MSH/MSE$. The F-test rejects H_0 when $\mathcal{F}$ is large. As outlined in the next section, the distribution of $\mathcal{F}$ is noncentral F with f_1 and f_2 degrees of freedom in the numerator and the denominator, respectively, and noncentrality parameter

$$\varphi = \sqrt{\frac{S^*}{f_1 + 1}}$$

where $S^* = ||E(\hat{\boldsymbol{\mu}} - \hat{\hat{\boldsymbol{\mu}}})||^2 = ||\boldsymbol{\mu} - E(\hat{\hat{\boldsymbol{\mu}}})||^2$. If H_0 is true, then $\varphi = 0$.

3.4 The Canonical Form

This section presupposes a basic background in matrix and vector algebra.

Consider the general linear model written in the form $Y_i = \mu_i + \epsilon_i$ where $\boldsymbol{\mu} = (\mu_1, \mu_2, \ldots, \mu_N) \in \boldsymbol{\Theta}^{(\boldsymbol{\mu})}$ and H_0: $\boldsymbol{\mu} \in \boldsymbol{\Theta}_0^{(\boldsymbol{\mu})}$ where $\boldsymbol{\Theta}^{(\boldsymbol{\mu})}$ is a $(k+1)$-dimensional subspace of $\mathcal{R}^N$ and $\boldsymbol{\Theta}_0^{(\boldsymbol{\mu})}$ is an $(l+1)$-dimensional subspace which is contained in $\boldsymbol{\Theta}^{(\boldsymbol{\mu})}$. Then, $\boldsymbol{\Theta}_0 = \boldsymbol{\Theta}_0^{(\boldsymbol{\mu})} \times (0, \infty)$.

There exists an orthogonal matrix $\boldsymbol{C}$ $(\boldsymbol{C}^{-1} = \boldsymbol{C}')$ such that $\boldsymbol{C\mu} = \boldsymbol{\nu} = (\nu_1, \nu_2, \ldots, \nu_N)$ where $\nu_{k+2} = \cdots = \nu_N = 0$ and, if H_0 is true, then $\nu_{l+2} = \cdots = \nu_{k+1} = 0$ as well. Furthermore, $\boldsymbol{CY} = \boldsymbol{Z} = (Z_1, Z_2, \ldots, Z_N)$ where $Z_1, \ldots, Z_N$ are independent and normally distributed with $E(Z_i) = \nu_i$ and $Var(Z_i) = \sigma^2$. Letting $\hat{\boldsymbol{\nu}} = \boldsymbol{C}\hat{\boldsymbol{\mu}}$ and $\hat{\hat{\boldsymbol{\nu}}} = \boldsymbol{C}\hat{\hat{\boldsymbol{\mu}}}$, the following relationships hold.

1. $\hat{\boldsymbol{\nu}}$ minimizes $||\boldsymbol{Z} - \boldsymbol{\nu}||^2$

2. $\hat{\hat{\boldsymbol{\nu}}}$ minimizes $||\boldsymbol{Z} - \boldsymbol{\nu}||^2$ when H_0 is true

3. $\hat{\nu}_i = \begin{cases} Z_i & \text{for } i = 1, \ldots, k+1 \\ 0 & \text{for } i = k+2, \ldots, N \end{cases}$

4. $\hat{\hat{\nu}}_i = \begin{cases} Z_i & \text{for } i = 1, \ldots, l+1 \\ 0 & \text{for } i = l+2, \ldots, N \end{cases}$

5. $SSE = ||\boldsymbol{Z} - \hat{\boldsymbol{\nu}}||^2 = \sum_{i=k+2}^{N} Z_i^2$ which is a chi-square random variable with $f_2 = N - (k+1)$ degrees of freedom

6. $SSH = ||\hat{\boldsymbol{\nu}} - \hat{\hat{\boldsymbol{\nu}}}||^2 = \sum_{i=l+2}^{k+1} Z_i^2$ which is a noncentral chi-square random variable with $f_1 = k - l$ degrees of freedom and noncentrality parameter

$$\varphi = \sqrt{\frac{S^*}{f_1 + 1}}$$

where

$$S^* = \sum_{i=l+2}^{k+1} \nu_i^2 \tag{3.3}$$

Hence, $\mathcal{F} = MSH/MSE$ has a noncentral F-distribution with $k - l$ and $N - k - 1$ degrees of freedom in the numerator and denominator, respectively.

It only remains to show that $S^* = ||\mu - E(\hat{\mu})||^2$. From (3.3)

$$S^* = \sum_{i=1}^{N} [E(\hat{\nu}_i - \hat{\nu}_i)]^2 = ||E(\hat{\nu} - \hat{\nu})||^2$$

$$= ||CE(\hat{\mu} - \hat{\mu})||^2 = ||E(\hat{\mu} - \hat{\mu})||^2$$

where the last equation follows from properties of orthogonal matrices.

Thus, S^* is obtained by substituting each Y_i by its expectation in any expression for SSH.

Since the Z_i in the expression for SSE all have mean 0, it follows that $E(MSE) = \sigma^2$. Furthermore, since $E(Z_i^2) = \sigma^2 + \nu_i^2$, it follows that

$$E(MSH) = \sigma^2 + \sum_{i=l+2}^{k+1} \nu_i^2/f_1 = \sigma^2 + S^*/f_1$$

3.5 Expressions for φ

In this section we summarize some of the expressions for φ for the models that have been discussed. In particular, the model and the null hypothesis H_0 are given, and the form for φ is listed.

(a) Model: General

$$\varphi = \sqrt{\frac{S^*}{\sigma^2(f_1 + 1)}}$$

where S^* = numerator sum of squares with each Y_i replaced by $E(Y_i)$, and f_1 = numerator degrees of freedom.

(b) Model: One-way layout; randomized blocks
H_0: Equal treatment means

$$\varphi = \sqrt{\frac{n \sum_i \delta_i^2}{I\sigma^2}}$$

where n = number of observations per treatment (number of blocks in randomized blocks), $\delta_i = \mu_i - \bar{\mu}$, μ_i = *i-th* treatment mean, and I = number of treatments.

(c) Model: Two-way layout (with or without interactions)

H_0: Zero row main effects

$$\varphi = \sqrt{\frac{Jn \sum_i \delta_i^2}{I\sigma^2}}$$

where n = number of observations per cell, δ_i = row main effect, $\delta_. = 0$, I = number of rows, and J = number of columns.

(d) Model: Two-way layout

H_0: Zero interactions

$$\varphi = \sqrt{\frac{n \sum_i \sum_j \rho_{ij}^2}{[(I-1)(J-1)+1]\sigma^2}}$$

where ρ_{ij} = interaction in cell (i, j). Other notation as in (c).

(e) Model: Latin squares

H_0: $\gamma_k = 0 \quad (k = 1, \ldots, m)$

$$\varphi = \sqrt{\frac{n \sum_k \gamma_k^2}{\sigma^2}}$$

where γ_k = effect of *k-th* level of factor C, $\gamma_. = 0$, n = number of levels of factors A and B, m = number of replicates.

(f) Model: Simple linear regression

H_0: $\delta_1 = 0$

$$\varphi = \sqrt{\frac{\delta_1^2(\sum_i x_i^2 - x_.^2/N)}{2\sigma^2}}$$

where $Y_i = \delta_0 + \delta_1 x_i + \epsilon_i \quad (i = 1, \ldots, N)$.

(g) Model: Quadratic regression

1. H_0: $\delta_2 = 0$

$$\varphi = \sqrt{\frac{\delta_2^2[\sum_i x_i^4 - (\sum_i x_i^2)^2/N]}{2\sigma^2}}$$

where $Y_i = \delta_0 + \delta_1 x_i + \delta_2 x_i^2 + \epsilon_i \quad (i = 1, \ldots, N), \quad x_. = \sum x_i^3 = 0.$

2. H_0: $\delta_1 = \delta_2 = 0$

$$\varphi = \sqrt{\frac{\delta_1^2 \sum_i x_i^2 + \delta_2^2[\sum_i x_i^4 - (\sum_i x_i^2)^2/N]}{3\sigma^2}}$$

(h) Model: Multivariate t, one-sample

H_0: $\boldsymbol{\mu} = \boldsymbol{\mu}_0$

$$\varphi = \sqrt{N(\boldsymbol{\mu} - \boldsymbol{\mu}_0)' \boldsymbol{\Sigma}^{-1}(\boldsymbol{\mu} - \boldsymbol{\mu}_0)/(p+1)}$$

where n = sample size, p = dimension, and $\boldsymbol{\mu}$ = mean vector.

(i) Model: Multivariate t, two-sample

H_0: $\boldsymbol{\mu}^{(1)} = \boldsymbol{\mu}^{(2)}$

$$\varphi = \sqrt{N_1 N_2(\boldsymbol{\mu}^{(1)} - \boldsymbol{\mu}^{(2)})' \boldsymbol{\Sigma}^{-1}(\boldsymbol{\mu}^{(1)} - \boldsymbol{\mu}^{(2)})/[(N_1 + N_2)(p+1)]}$$

where $\boldsymbol{\mu}^{(i)}$ = i-th treatment mean, $N = N_1 + N_2$, N_i = i-th treatment sample size, and p = dimension.

(j) Model: Profile analysis, two-sample

H_0: $\mu_i^{(1)} - \mu_{i+1}^{(1)} = \mu_i^{(2)} - \mu_{i+1}^{(2)} \quad (i = 1, \ldots, p-1)$

$$\varphi = \sqrt{\frac{N_1 N_2}{p(N_1 + N_2)}(\boldsymbol{\mu}^{(1)} - \boldsymbol{\mu}^{(2)})'\boldsymbol{C}'(\boldsymbol{CSC}')^{-1}\boldsymbol{C}(\boldsymbol{\mu}^{(1)} - \boldsymbol{\mu}^{(2)})}$$

where

$$\boldsymbol{C} = \begin{bmatrix} 1 & -1 & 0 & 0 & 0 & \ldots & 0 & 0 & 0 \\ 0 & 1 & -1 & 0 & 0 & \ldots & 0 & 0 & 0 \\ \vdots & \vdots & \vdots & \vdots & \vdots & & \vdots & \vdots & \vdots \\ 0 & 0 & 0 & 0 & 0 & \ldots & 0 & 1 & -1 \end{bmatrix}$$

Other notation as in (i).

Chapter 4

LITERATURE SURVEY

This chapter is concerned with other forms in which data connected with the operating characteristic of the F-test have been presented. For some cases special purpose charts exist which are easier to use than ours. It should be kept in mind that our charts have as their purpose easy calculation of the the sample size needed to achieve a desired power. The linear models are general, and the ranges of α and power are very wide.

4.1 Tables of the Operating Characteristic

The first tables dealing with the power of the F-test were prepared by Tang (1938). He tabled the operating characteristic (1-power) for given α, f_1, f_2, and φ. Extensions of his tables have been made by Tiku (1956, 1972) and Lachenbruch (1967).

The ranges of values for the parameters in these tables are as follows:

Tang:	$\alpha = .01, .05$
	$f_1 = 1(1)8$
	$f_2 = 2, 4, 6(1)30, 60, \infty$
	$\varphi = 1.(.5)3.(1.)8.0$
Tiku:	$\alpha = .005, .01, .025, .05$ (1967), and .1 (1972)
	$f_1 = 1(1)10, 12$
	$f_2 = 2(2)30, 40, 60, 120, \infty$
	$\varphi = .5, 1.(.2)2.2(.4)3.0$

Lachenbruch: $\alpha = .01, .05$
$f_1 = 1(1)12, 14, 16, 20, 24, 30, 40, 50, 75$
$f_2 = 2(2)20(4)40(10)80$
$\varphi = 1.(.5)3.(1.)8.0$

Tang has also tabled E^2_α, related to the upper tail cutoff point F_α of the central F-distribution by

$$F_\alpha = \frac{f_2}{f_1}\frac{E^2_\alpha}{1 - E^2_\alpha}$$

Lachenbruch included tables of the percentage points of the noncentral F-distribution and a FORTRAN program for their computation. In place of φ, he used the parameter

$$\lambda = \frac{(f_1 + 1)\varphi^2}{2} = \frac{S^*}{2\sigma^2}$$

These tables enable one to obtain the power function and, by trial and error coupled with interpolation, to solve for sample size. The range of parameter values, especially of φ, is too limited for some purposes.

Ifram (1971) tabled $P(X \leq x)$, the cumulative distribution function, when X has a noncentral β-distribution. If the usual F-ratio, F', has the same noncentrality parameter, then the operating characteristic is

$$P(F' \leq F_\alpha) = P\left(X \leq \frac{f_1 F_\alpha}{f_1 F_\alpha + f_2}\right)$$

The parameter ranges are f_1, $f_2 = 1(1)5$, $f_1 + f_2 \leq 6$, $x = .05(.05).95$, and $\theta = 0.(1.)10.0$ where

$$\theta = (f_1 + 1)\varphi^2 = \frac{S^*}{\sigma^2}$$

The range of parameter values is too limited for most sample size computations.

4.2 Tables of φ

For $\alpha = .01, .05$; $f_1 = 1(1)10$, 12, 15, 20, 24, 30, 40, 60, 120, ∞; and $f_2 = 2(2)20$, 24, 30, 40, 60, 120, 240, ∞, Lehmer (1944) tabled the values of φ which give powers .7 and .8. For sample size determination these tables involve interpolation and cover too limited values of the power.

An extension of Lehmer's tables has been prepared by the National Bureau of Standards (undated), but will not be published. Ranges covered are $\alpha = .01$, .02, .05, .10, and .20; power (denoted by β) .10, .50, .90, .95, and .99; $f_1 = 1(1)10$, 12, 15, 20, 24, 30, 40, 60, 120, ∞ (as in Lehmer's tables); and $f_2 = 2(2)12$, 20, 24, 30, 40, 60, ∞.

Similar tables were presented by Dasgupta (1968). His range is $\alpha = .01, .05$; power = .1(.1).9; f_1 (denoted by M)= 1(1)10; and f_2 (denoted by N) = 10(5)50(10) 100, ∞. Instead of φ he tabled values of λ (denoted by ∂).

4.3 Curves of Constant f_1 or f_2

For $\alpha = .01, .05$ and $f_1 = 1(1)8$, Pearson and Hartley (1951, 1972) plotted curves of constant f_2 = 6(1)10, 12, 15, 20, 30, 60, ∞. The coordinate scales are φ and power. These charts are reproduced in Scheffé (1959). Duncan (1957) condensed the Pearson and Hartley charts by limiting power to .5 and .9 and replacing the power coordinate by f_1. Other parameter ranges are identical.

For their limited range these charts are reasonably usable for choice of sample size.

Patniak (1949) plotted values of $\lambda = S^*/\sigma^2$ against f_2 for $\alpha = .05$, power = .5, .9 and $f_1 = 1(1)10$. He denoted f_1 and f_2 by ν_1 and ν_2, respectively, and the power by $\beta(\nu_1, \nu_2, \lambda)$.

4.4 Computer Programs

Bargmann and Ghosh (1964) have written FORTRAN programs for the cumulative distribution function for noncentral F (operating characteristic of the $F-$test) and for the percentage points of the noncentral $F-$distribution. With additional programming, the latter can be used in a trial and error search for sample size.

In addition, similar programs are provided for the noncentral χ^2-, $\beta-$, and $t-$distribution. In an earlier paper, Bargmann and Ghosh (1963) had given similar programs for the central versions of the above distributions as well as for the standard normal and $\gamma-$distribution. The programs for cumulative distribution functions also compute ordinates of the density functions.

Some of the programs given by Bargmann and Ghosh have been updated by Bargmann and Thomas (1971) and by Bouver and Bargmann (1975).

4.5 Special Purpose Charts and Tables

Keuls (1960a-c) has prepared charts and tables in several forms giving relationships between the number of treatments, power, number of replicates, the noncentrality parameter, and σ^2 for randomized block experiments. He uses

$$A = \sqrt{\sigma^2\varphi^2\frac{(f_1+1)}{r}} = \sqrt{\frac{S^*}{r}}$$

instead of φ. Here r is the number of replicates. He set $n = f_1 + 1$, the number of treatments. The most relevant of these are

1. A table of the Lehmer type giving values of $\gamma = A\sqrt{r/\sigma} = 30, 40, 60$; r =2(1)6, 8, 10, 20; and $\alpha = .05$ (1960a).

2. Nomograms giving approximate relationships between A, σ, n, r, and power for $\alpha = .05$ (1960a).

3. Graphs for finding r when A, σ, and n are given with $\alpha = .05, .1$, and power .5, .9 (1960a-c).

For the one-way layout analysis of variance, Feldt and Mahmoud's (1958) charts give n, the number of replicates needed to yield power .5, .7, .8, .9, .95 when $\alpha = .01$, .05, and the number of treatments is 2(1)5. In place of φ they use

$$\lambda = \frac{\varphi}{\sqrt{n}}$$

All of the above special purpose presentations can, with some effort, be used in more general cases.

Asaki, Kondo, and Narita (1970) considered testing for main effects with $\alpha =$ .05. For a one-way layout they provide charts for power .8 and .9. Let δ be the maximum absolute difference between means and $k = \delta/\sigma$. For given k, the number of replicates varies between a minimum (when half the means are at each extreme) and a maximum (when the nonextreme means are all halfway between the extremes). Their charts give the minimum and maximum numbers of replications as the number of levels varies in the range 2(1)10. An indication is given of how the charts can be adapted to higher order analysis or variances.

Similar charts for interactions in two-way layouts were given by Asaki and Kondo (1971). Again, adaptation to higher order analysis of variance is indicated.

For the one-way layout a more extensive table of the type above has been given by Bratcher, et. al. (1970). They only gave the maximum number of replicates, but their parameters vary over $\alpha = .01, .05, .1, .2, .25, .3, .4, .5$; $\delta/\sigma = 1.(.25)2., 2.5, 3.$ (denoted by Δ/σ); power = .7, .8, .9, .95; and number of levels = 2(1)11, 13, 16, 21, 25, 31.

Bowman and Kastenbaum (1975) prepared tables appropriate to the one- and two-way layouts. Both set of tables are entered with α =.01, .05, .1, .2; and β (1-power) = .005, .01, .025, .05, .1(.1).4.

In the one-way layout (Table I) K is the number of groups to be compared, and N is the number of observations in each of the K groups. For given values of α, β, K, and N, the tables give the maximum value τ of the standardized range of the the group means which yields the desired power. The relationship between τ and φ is

$$\tau = \varphi\sqrt{\frac{2K}{N}}$$

attained when all but the two extreme cell means are at the average of the extremes. Values of τ are given for $N = 2(1)30(5)50(10)100(50)200(100)500$, 1000; and $K = 2(1)11(2)15(5)30(10)60$.

In the two-way layout (Table II) K is the number of treatments, B is the number of blocks, and N is the number of observations per cell. The value of τ, in this case the standardized range of the treatment means, is tabled for $N = 1(1)5$; $B = 2(1)5$, and values of K as given in the above description of Table I. The relationship between τ and φ is given by

$$\tau = \varphi\sqrt{\frac{2K}{NB}}$$

These tables supersede the tables of Kastenbaum, et. al. (1970a-b) and have the advantage that no iteration is required. The authors also show how the tables may be used to find sample sizes for the three-way layout and Latin Squares designs.

4.6 Stein's Two-sample Procedure

One difficulty with the form of the power of the usual F−test is the dependence of φ in Eq. (1.4) on σ^2. Thus, deviations from the hypothesis are measured in units of σ, a disadvantage in many applications.

Danzig (1940) showed that in Student's case ($f_1 = 1$) no fixed sample test exists with power independent of σ^2. This result was extended to arbitrary f_1 by Stein (1945). Stein also gave a two-sample procedure with power independent of σ^2. For this procedure, the total sample size, while finite, is unbounded.

To apply Stein's procedure, one must evaluate a quantity which depends on the desired power, the model and the alternative. Rodger (1976) has provided tables of this quantity for $\alpha = .01$, .05 and powers .5, .7, .8, .9, .95, and .99.

4.7 Error Rate

Rodger(1974) proposed a post-hoc method by which a set of f_1 orthogonal contrasts among the means in a one-way layout can be examined. Consider H_{i0}: the i-th contrast is 0 vs H_{i1}: the i-th contrast is nonzero. By his method, one guarantees that at most r of the H_{i0} will be rejected. Rodger defines the *Type I error rate* to be the expected proportion of errors when all the H_{i0} are true and the *power* to be the expected proportion of rejections when all the H_{i0} are false. Tables of critical values for the test were given (Rodger 1975a) for α =.01, .02; f_1 =1(1)20(10)60; and f_2 = 4(2)30(10)60, ∞. For the same values of α, f_1, and f_2, Rodger (1975b) gave tables to allow the computation of sample sizes for power β = .5, .7, .8, .9, .95, .99.

Rodger(1978) gave a Stein type two sample procedure applied to his error rate approach. Tables analogous to his earlier (1976) tables are given.

Chapter 5

DETAILS OF COMPUTATION

In this chapter we describe in detail the methods used in constructing the tables and charts. Sec. 5.1 describes the numerical methods used to evaluate the incomplete γ-function ratio and the incomplete β-function ratio. It is also shown how percentage points of the functions are obtained. These results are also used in Sec. 5.2 to obtain critical values of the $\chi^2, F-$ and $t-$distributions. The results are also used in Sec. 5.3, where numerical methods are described for obtaining the power of the $F-$test.

The methods used to construct the charts are discussed in detail in Sec. 5.4.

5.1 Auxiliary Functions

The following functions are used both in the computation of the the tables of percentage points and in the computation of the power of the $F-$test.

5.1.1 The incomplete $\gamma-$function ratio.

The incomplete $\gamma-$function ratio is defined as

$$I(x,s) = \frac{1}{\Gamma(s)} \int_0^x \exp(-t)\, t^{s-1}\, dt \quad \text{for } x \geq 0,\ s > 0. \tag{5.1}$$

For given values of $s \leq x \leq 1$ or for $x < s$, this function was evaluated using the series expansion

$$I(x,s) = \frac{\exp(-x)x^s}{\Gamma(s+1)} \Big[1 + \sum_{r=1}^{\infty} \frac{x^r}{(s+1)(s+2)\cdots(s+r)}\Big] \tag{5.2}$$

For all other cases the function was evaluated using the continued fraction expansion

$$I(x,s) = \frac{\exp(-x)x^s}{\Gamma(s)} \Big(\frac{1}{x+}\,\frac{1-s}{1+}\,\frac{1}{x+}\,\frac{2-s}{1+}\cdots\Big) \tag{5.3}$$

The FORTRAN program given by Bhattacharjee (1970) was used to evaluate $I(x, s)$ by (5.2) and (5.3)

The accuracy of the routine was set at 10^{-12}. Thus, the series in (5.2) was continued until the last term computed was less than 10^{-12}, and the continued fraction in (5.3) was continued until the relative error between two successive convergents was less than 10^{-12}.

If x'_p is defined to be the upper *100p-th* percentage point of $I(x, s)$, then

$$I(x'_p, s) = 1 - p \tag{5.4}$$

To obtain x'_p, Newton-Raphson iteration was used with starting value

$$x'_p \approx \frac{s}{2}\left(1 - \frac{2}{9s} + y_p\sqrt{\frac{2}{9s}}\right)^3 \tag{5.5}$$

where y_p is the upper percentage point of the standard normal distribution. (See Abramowitz and Stegun, 1968, Eq. 26.4.17, pp. 941).

The iteration was continued until the difference between two successive approximations was less than 10^{-5}.

An approximation to y_p (see Odeh and Evans, 1972 and 1974) is given by

$$y_p = t - \frac{P_4(t)}{q_4(t)} \tag{5.6}$$

where $t = \sqrt{-2\ln p}$, $P_4(t) = \sum_{i=0}^{4} a_i t^i$, and $Q_4(t) = \sum_{j=0}^{4} b_j t^j$. The coefficients of $P_4(t)$ and $Q_4(t)$ are:

$a_0 = .32223\ 24310\ 87751$ $\quad$ $b_0 = .99348\ 46260\ 59901 \times 10^{-1}$

$a_1 = 1.0$ $\quad$ $b_1 = .58858\ 15704\ 94926$

$a_2 = .34224\ 20885\ 47343$ $\quad$ $b_2 = .53110\ 34623\ 66099$

$a_3 = .20423\ 12102\ 45003 \times 10^{-1}$ $\quad$ $b_3 = .10353\ 77528\ 49707$

$a_4 = .45364\ 22101\ 47524 \times 10^{-4}$ $\quad$ $b_4 = .38560\ 70063\ 39815 \times 10^{-2}$

The maximum error in the approximation given by (5.6) is 1.6×10^{-8} for values of p in the range $10^{-20} \leq p \leq .5$. For values of $p > .5$, the value of y_p was found from the relationship $y_{1-p} = -y_p$.

5.1.2 The incomplete β–function ratio

The incomplete β–function ratio is defined as

$$I_x(a,b) = \frac{1}{B(a,b)} \int_0^x t^{a-1}(1-t)^{b-1}\,dt \tag{5.7}$$

$$a > 0,\ b > 0,\ 0 \le x \le 1.$$

For given values of a and b, and for $0 < x < 1$, this function was evaluated using the continued fraction expansion (see Abramowitz and Stegun, 1968, Eq. 26.5.9, pp. 944).

$$I_x(a,b) = \frac{x^a(1-x)^{b-1}}{aB(a,b)}\left(\frac{c_1}{1+}\frac{c_2}{1+}\frac{c_3}{1+}\cdots\right) \tag{5.8}$$

where $c_1 = 1$,

$$c_{2m} = \frac{(a+m-1)(b-m)}{(a+2m-2)(a+2m-1)} \cdot \frac{x}{1-x}$$

and

$$c_{2m+1} = \frac{m(a+b-1+m)}{(a+2m-1)(a+2m)} \cdot \frac{x}{1-x}$$

The series was continued until the relative error in two successive even convergents was less than 10^{-12}.

If the lower percentage point of $I_x(a,b)$ is x_p then

$$I_{x_p}(a,b) = p \tag{5.9}$$

To find x_p the Newton-Raphson iteration formula was used (see Abramowitz and Stegun, 1968, Eq. 26.5.22, pp. 945) with starting value

$$x_p \approx \frac{a}{a+be^{w}} \tag{5.10}$$

where

$$w = \frac{y_p(h+\lambda)^{1/2}}{h} - \frac{2}{h}\left(\lambda + \frac{5}{6} - \frac{2}{3h}\right)$$

$$h = 2\left(\frac{1}{2a-1} + \frac{1}{2b-1}\right)^{-1} \quad \text{if } a \ne \frac{1}{2},\ b \ne \frac{1}{2}$$

$$\lambda = \frac{y_p^2 - 3}{6}$$

$$w = 0 \text{ if } a = \frac{1}{2} \text{ or } b = \frac{1}{2}$$

and y_p is defined by (5.6).

5.2 Generation of Critical Values

5.2.1 The χ^2-distribution

Define $\chi^2(\nu)$ to be a random variable having a χ^2-distribution with parameter ν. The critical value $\chi^2_\alpha(\nu)$ is defined by

$$P[\chi^2(\nu) > \chi^2_\alpha(\nu)] = \alpha \tag{5.11}$$

Since $\chi^2(\nu)/2$ has a γ-distribution with parameter $\nu/2$, it follows that

$$P[\chi^2(\nu)/2 \leq x] = I(x, \nu/2) \tag{5.12}$$

where $I(x, s)$ is defined by (5.1).

From the above and from (5.4) it follows that

$$P[\chi^2(\nu)/2 \leq \chi^2_\alpha(\nu)/2] = I(x'_{1-\alpha}, \nu/2) = 1 - \alpha \tag{5.13}$$

and thus,

$$\chi^2_\alpha(\nu) = 2x'_{1-\alpha} \tag{5.14}$$

An efficient computer algorithm for generating $\chi^2_\alpha(\nu)$ is given by Goldstein (1973).

5.2.2 The F-distribution

The critical value $F_\alpha(f_1, f_2)$ of the variance ratio F with f_1 and f_2 degrees of freedom satisfies

$$P[F > F_\alpha] = \alpha \tag{5.15}$$

and is given by

$$F_\alpha = \frac{f_2}{f_1} \cdot \frac{1 - x_\alpha}{x_\alpha} \tag{5.16}$$

where x_α satisfies $I_{x_\alpha}(f_2/2, f_1/2) = \alpha$ and $I_x(a, b)$ is defined by (5.9).

If $f_2 = \infty$, then F is distributed $\chi^2(f_1)/f_1$. Thus,

$$\alpha = P[F > F_\alpha] = P[\chi^2(f_1)/f_1 > F_\alpha] = P[\chi^2(f_1)/f_1 > \chi^2_\alpha(f_1)/f_1]$$

so that

$$F_\alpha = \frac{\chi^2_\alpha(f_1)}{f_1} \tag{5.17}$$

where $\chi^2_\alpha(f_1)$ is defined from Sec. 5.2.1.

If $f_1 = \infty$, then the variance ratio F is distributed as $f_2/\chi^2(f_2)$. It follows that

$$\alpha = \mathrm{P}[F > F_\alpha] = \mathrm{P}[f_2/\chi^2(f_2) > F_\alpha]$$

$$= \mathrm{P}[\chi^2(f_2) \le f_2/F_\alpha] = \mathrm{P}[\chi^2(f_2) \le \chi^2_{1-\alpha}(f_2)]$$

and, thus

$$F_\alpha = \frac{f_2}{\chi^2_{1-\alpha}(f_2)} \tag{5.18}$$

5.2.3 The t-distribution

The critical value $t_\alpha(\nu)$ of the t-distribution with ν degrees of freedom satisfies

$$\mathrm{P}[t(\nu) > t_\alpha(\nu)] = \alpha \tag{5.19}$$

If $f_1 = 1,\ f_2 = \nu$, the variance ratio F is distributed as t^2. Thus

$$t_\alpha(\nu) = \sqrt{F_{2\alpha}(1,\nu)} \tag{5.20}$$

If $\nu = \infty$, then $t_\alpha(\infty) = y_\alpha$, where y_α is given by (5.6). Accurate approximations to $t_\alpha(\nu)$ are given by the computer algorithm of Hill(1970).

5.3 Power Computation

5.3.1 Finite f_1 and f_2

Let F' denote a random variable having a noncentral F–distribution with f_1 and f_2 degrees of freedom, and noncentrality parameter λ. For a given value of f_1, f_2, α, and φ, the power of the F–test is given by

$$\text{Power} = \mathrm{P}[F' > F_\alpha|\lambda]$$

where $\lambda = (1 + f_1)\varphi^2$.

If f_2 is an even integer, the power is evaluated using the finite series (see Abramowitz and Stegun, 1968, Eq. 26.6.22, pp. 948):

$$\mathrm{P}[F' < F_\alpha|\lambda] = e^{-\lambda(1-x_\alpha)/2}\, x_\alpha^{(f_1+f_2-2)/2} \sum_{i=0}^{f_2/2-1} T_i \tag{5.21}$$

where

$$T_0 = 1$$

$$T_1 = \frac{1}{2}(f_1 + f_2 - 2 + \lambda x_\alpha) \cdot \frac{1 - x_\alpha}{x_\alpha}$$

$$T_i = \frac{1}{2i}[(f_1 + f_2 - 2i + \lambda x_\alpha)T_{i-1} + \lambda(1 - x_\alpha)T_{i-2}] \cdot \frac{1 - x_\alpha}{x_\alpha}$$

and

$$I_{x_\alpha}(f_2/2, f_1/2) = \alpha$$

For any value of f_2, a three moment noncentral F approximation was developed by Tiku (1965) and examined by him (1966). The approximation is

$$\text{(5.22)} \qquad \mathrm{P}[F' > F_\alpha|\lambda] \approx I_{y_0}(f_2/2, b/2)$$

where

$$H = 2(f_1 + \lambda)^3 + 3(f_1 + 2\lambda) + (f_1 + 3\lambda)(f_2 - 2)^2$$

$$K = (f_1 + \lambda)^2 + (f_2 - 2)(f_1 + 2\lambda)$$

$$E = \frac{H^2}{K^3}$$

$$b = \frac{1}{2}(f_2 - 2)\left\{\left(\frac{F}{E - 4}\right)^{1/2} - 1\right\}$$

$$h = \frac{b}{f_1} \frac{1}{(2b + f_2 - 2)} \frac{H}{K}$$

$$C = \frac{f_2}{f_2 - 2}\left\{h - \frac{f_1 + \lambda}{f_1}\right\}$$

and

$$y_0 = \frac{h}{[h + (\frac{b}{f_2})(F_\alpha + C)]}$$

5.3.2 Infinite f_1 and f_2

AS f_1 tends to infinity, $\mathrm{P}[F' > F_\alpha|\lambda]$ has the limiting form (See Abramowitz and Stegun (1968), Eq. 26.6.24, pp. 948):

$$\text{(5.23)} \qquad \lim_{f_1 \to \infty} \mathrm{P}[F' > F_\alpha|\lambda] = \mathrm{P}\left[\chi^2(f_2) \leq \frac{f_2(1 + c^2)}{F_\alpha(\infty, f_2)}\right]$$

where

$$\lim_{f_1 \to \infty} \frac{\lambda}{f_1} = c^2$$

and

$$F_\alpha(\infty, f_2) = \frac{f_2}{\chi^2_{1-\alpha}(f_2)}$$

Since $\lambda = (1 - f_1)\varphi^2$, then $c^2 = \varphi^2$ and (5.23) becomes

$$\lim_{f_1 \to \infty} \mathrm{P}[F' > F_\alpha | \lambda] = \mathrm{P}[\chi^2(f_2) \leq (1 + \varphi^2)\chi^2_{1-\alpha}(f_2)]$$

$$= \mathrm{P}[\chi^2(f_2)/2 \leq (1 + \varphi^2)\chi^2_{1-\alpha}(f_2)/2] \tag{5.24}$$

$$= I\Big((1 + \varphi^2)\chi^2_{1-\alpha}(f_2)/2, f_2/2\Big)$$

where $I(x, s)$ is defined by (5.1).

As f_2 tends to infinity, $\mathrm{P}[F' > F|\lambda]$ has the limiting form (see Abramowitz and Stegun, 1968, Eq. 26.6.23, pp. 942):

$$\lim_{f_2 \to \infty} \mathrm{P}[F' > F_\alpha | \lambda] = \mathrm{P}[\chi'^2(f_1) > \chi^2_\alpha(f_1)|\lambda] \tag{5.25}$$

where $\chi'^2(f_1)$ is a random variable having a noncentral χ^2–distribution with f_1 degrees of freedom and noncentrality parameter $\lambda = f_1(1 + \varphi^2)$. Then $\mathrm{P}[\chi'^2(f_1) > \chi^2_\alpha(f_1)|\lambda]$ was evaluated using the series:

$$1 - e^{-\lambda/2} \sum_{j=0}^{\infty} \frac{(\lambda/2)^j}{j!} \mathrm{P}\big[\chi^2(f_1 + 2j) \leq \chi^2_\alpha(f_1)\big] \tag{5.26}$$

where

$$\mathrm{P}\big[\chi^2(f_1 + 2j) \leq \chi^2_\alpha(f_1)\big] = I\big(\chi^2_\alpha(f_1)/2, f_1/2 + j\big)$$

and where $I(x, s)$ is defined by (5.1) (see Abramowitz and Stegun, 1968, Eq. 26.4.25, pp. 942).

The series given in (5.26) was continued until the error of the last term computed relative to the partial sum was less than 10^{-4}.

5.4 Some More Recent Algorithms

Many relevant algorithms have appeared in the literature since we first did the computation for these charts in 1974. In this section we list some of them. The list given here is not meant to be exhaustive but rather to point to some readily available algorithms.

An algorithm for computing the incomplete γ-function ratio is given by Lau (1980).

Beasley and Springer (1985) give an algorithm for finding percentage points of the normal distribution.

An algorithm for computing the incomplete β-function ratio is given by Majumder and Bhattacharjee (1985a). These authors also give an algorithm for finding percentage points (1985b).

Best and Roberts (1985) give an algorithm for finding percentage points of the χ^2 distribution.

Narula and Desu (1981) give algorithms for computing the cumulative of the noncentral χ^2 distribution and also for finding the noncentrality parameter.

Narula and Weistroffer (1986) give algorithms for computing the cumulative of the noncentral F-distribution and also for finding the noncentrality parameter.

Other algorithms for computing the cumulative of the noncentral F-distribution are given by Norton (1983) and by Lenth (1983).

5.5 Construction of the Charts

5.4.1 General details

The charts were drawn on a plotter from a tape generated by a computer program. For $f_1 \geq 3$, $f_2 \geq 3.8$, the scales on the charts are reciprocal. The original charts were drawn so that the horizontal distance between $f_1 = 2$ and $f_2 = 3$ is 2.5 inches. The horizontal distance between $f_1 = 2$ and $f_1 = \infty$ is 10 inches. The vertical scale for f_2 is the same as the horizontal scale for f_1. The horizontal scale of the left-hand box was chosen so that the horizontal distance between $f_1 = 1$ and $f_1 = 2$ was 1.25 inches.

The minimum value of φ which appears on a particular chart and the value by which φ was incremented were determined by trial and error. The curve in the upper right-hand corner of the chart was computed first and then φ was incremented. The computer program generated curves until a value of φ was reached for which no part of the line would appear on the chart; that is, f_2 was always less than 3.8 for all values of f_1. The curves of constant φ which appear in the left-hand box are those lines which intersect the line $f_2 = \infty$ or $f_2 = 3.8$ for a value of $f_1 < 1$.

For a given value of α and power, each curve of constant φ is constructed in the following way: for fixed f_1 we determine the value of f_2 which gives the required power. As f_1 is varied, a sequence of points on the curve is generated. The endpoints are determined separately, i.e., the values of f_1 for which $f_2 = \infty$ or $f_2 = 3.8$. Any two consecutive points are joined by straight lines to give the entire line of constant φ. If for $f_1 = \infty$, $f_2 > 80$, then only the two endpoints are joined. Points are determined for values of $f_1 = 1(1)10(2)20(20)100, 200, \infty$.

5.5.2 Numerical methods

For a fixed value of α and power, and a given value of f_1 and φ, the value of f_2 is found using one of the following two methods:

Method A. (See McCracken and Dorn, 1964 pp. 130). Let X_n denote the *n-th* approximation to the desired value of f_2, and let $P(X_n)$ denote the corresponding power. Denote by DP the desired power.

Then the *(n+1)-th* approximation to f_2 is given by

$$X_{n+1} = X_n + \frac{X_n - X_{n-1}}{P(X_n) - P(X_{n-1})}[DP - P(X_n)]. \tag{5.27}$$

This iteration was continued until $|X_{n+1} - X_n| < 10^{-4}$.

This method was used when $f_1 = \infty$ and $P(X_n)$ was computed from (5.24). It was also used for finite f_1 and $f_2 < 20$ and $P(X_n)$ was computed from Tiku's three-moment approximation given in (5.22).

Initial values of X_0 and X_1 were chosen in the following way: X_0 was chosen as the largest integer from the set $S = \{3, 6, 9, 19, 39\}$ such that $P(X_0) \leq DP$. (Note that if $P(3) > DP$, the line of constant φ does not intersect the line $f_1 = \infty$ on the chart.) X_1 was then chosen as $X_0 + 1$.

The above method was also used to find the value of f_1 where the line of constant φ intersects $f_2 = \infty$. In this case, X_n is the *n-th* approximation to f_1 and $P(X_n)$ was computed from (5.25). Here the starting values $X_0 = 3, X_1 = 4$ were used.

Method B. If f_1 is finite, but $f_2 > 20$, (5.21) was used to compute the power for the two even integers which surround the correct value of f_2. Linear interpolation was then used to find f_2. If $P(f_2) = DP$, denote by f_2^* the largest even integer which is less than or equal to f_2. Then

$$P(f_2^*) \leq DP \leq P(f_2^* + 2) \tag{5.28}$$

and f_2 is approximated by

$$f_2 \approx f_2^* + 2\frac{DP - P(f_2^*)}{P(f_2^* + 2) - P(f_2^*)}.$$

This combination of methods was used for the following reasons. For finite f_1 and $f_2 > 20$, method B will find f_2 to at least one decimal place. The three-moment approximation is not sufficiently accurate, particularly for power $\geq .8$ or $\alpha \geq .25$, to yield this accuracy. For $f_2 \geq 20$, the three-moment approximation is sufficiently accurate to yield two decimal places for f_2, and method A was used. Since this method is iterative, it is easy to use. Method B will not yield two decimal place accuracy. However, interpolation through four successive even values of f_2 gave agreement with method A to two decimal places.

Chapter 6

INTERPOLATION IN TABLE 1

(Critical Values of the F-distribution)

The critical values of the F-distribution are approximately linear in $1/f_1$ and $1/f_2$ provided that $f_1 > 12$ or $f_2 > 40$. This makes interpolation in Table 1 very simple.

Suppose that Table 1 provides an entry for the desired value of f_1 but not for f_2. Suppose that f_{21} and f_{22} are adjacent entry values of f_2 satisfying $f_{21} < f_2 < f_{22}$. Let the α critical value for f_1 and f_{21} degrees of freedom be a and for f_1 and f_{22} be b. Then the interpolated critical value for f_1 and f_2 degrees of freedom is

$$F_\alpha = b + \frac{(1/f_2 - 1/f_{22})}{(1/f_{21} - 1/f_{22})}(a - b). \tag{6.1}$$

Note that $1/\infty = 0$.

This will be illustrated by finding the critical value for Example 2.3.3, case 3. In that example $f_1 = 2$, an available entry, but $f_2 = 54$ which is unavailable. Then $f_{21} = 50$ and $f_{22} = 60$ so that from Table 1.8 in Part Two, $a = 7.956$ and $b = 7.768$ for $\alpha = .001$. Thus, (6.1) yields

$$F_{.001}(2, 54) = 7.768 + \frac{(1/54 - 1/60)}{(1/50 - 1/60)}(7.956 - 7.768) = 7.872$$

In the rare case that f_1 is not a tabled entry, but f_2 is, (6.1) with the obvious interchange is still valid.

Large value of f_1 are rare, but we will illustrate by example how repeated applications of (6.1) can be used to interpolate in both f_1 and f_2. Suppose that $\alpha = .05, f_1 = 65$ and $f_2 = 85$. First fix $f_1 = 60$. Then $f_{21} = 80$ and $f_{22} = 90$. Use Table 1.4 and (6.1) to obtain for $f_1 = 60$ and $f_2 = 85$ the result

$$F_{.05}(60,85) = 1.465 + \frac{(1/85 - 1/90)}{(1/80 - 1/90)}(1.482 - 1.465) = 1.473$$

Repeat this for $f_1 = 90$ and $f_2 = 85$ to obtain

$$F_{.05}(90,85) = 1.417 + \frac{(1/85 - 1/90)}{(1/80 - 1/90)}(1.436 - 1.417) = 1.426$$

Now use the obvious interchange in (6.1) for $f_1 = 65$ and $f_2 = 85$ to obtain the desired result

$$F_{.05}(65,85) = 1.426 + \frac{(1/65 - 1/90)}{(1/60 - 1/90)}(1.473 - 1.426) = 1.462$$

All of the interpolated values above agree with the exact answer. In general the error in interpolation will be at most one unit in the last decimal place.

For values of $f_1 \leq 12$ and $f_2 \leq 40$ linear interpolation in f_2 will also yield this accuracy.

REFERENCES

Examples

Beale, H.P. (1934). The serum reactions as an aid in the study of filterable viruses of plants. *Boyce Thompson Inst. for Plant Res., Contrib.* **6** 407-435.

Bliss, C.I. (1952). *The Statistics of Bioassay.* (New York: Academic Press, Inc.).

Burt, C. and Lewis, R.B. (1946). Teaching backward readers. *Br. J. Educ. Psych.* **16** 116-132.

Crampton, E.W. (1947). The growth of the odontoblasts of the incisor tooth as a criterion of the vitamin C intake of the guinea pig. *J. Nutrition* **33** 491-504.

Erdman, L.W. (1946). Studies to determine if antibiosis occurs among Rhizobia: 1. Between Rhizobium meliloti and Rhizobium trifolii. *J. Amer. Soc. Agron.* **38** 251-258.

Fisher, R.A. (1936). The use of multiple measurements in taxonomic problems. *Ann. Eugenics* **7** 179-188.

Glick, S.D. and Greenstein, S. (1973). Comparative learning and memory deficits following hippocampal and candidate lesions in mice. *J. Comp. and Physio. Psych.* **82** 188-194.

Hardison, W.A. and Reid, J.T. (1953). Use of indicators in the measurement of the dry matter intake of grazing animals. *J. Nutrition* **51** 35-52.

Kasschau, R.A. (1972). Polarization in serial and paired-associate learning. *Amer. J. Psych.* **85** 43-56.

Mason, K.E. (1942). Criteria of response in the bioassay of vitamin E. *J. Nutrition* **23** 59-70.

Neter, J., Wasserman, W. and Kutner, M.H. (1985). *Applied Linear Statistical Models*, 2nd edition. (Homewood, Ill. : Richard D. Irwin, Inc.)

Ostle, B. (1963). *Statistics in Research* (Ames: Iowa University Press).

Preuss, P.W., et. al. (1968). Studies of fluoro-organic compounds in plants. I. Metabolism of $2^{14}-$C-fluoroacetate. *Boyce Thompson Inst. for Plant Res., Contrib.* **24** 25-31.

Schroeder, E.M. (1945). *On Measurement of Motor Skills.* (New York: King's Crown Press).

Steel, R.G.D. and Torrie, J.H. (1960). *Principles and Procedures of Statistics with Special Reference to Biological Sciences.* (New York: McGraw-Hill).

Valenzi, E.R. and Andrews, I.R. (1971). Effect of hourly overpay and underpay inequity when tested with a new induction procedure. *J. Appl. Psych.* **55** 22-27.

Walker, H.M. and Lev J. (1953). *Statistical Inference.* (New York: Holt, Rinehart and Winston, Inc.).

Youden, W.J. and Beale, H.P. (1934). Statistical study of the local lesion method for estimating tobacco mosaic virus. *Boyce Thompson Inst. for Plant Res., Contrib.* **6** 437-454.

Theoretical References

Dantzig, G.B. (1940). On the non-existence of test of "Student's" hypothesis having power functions independent of σ. *Ann. Math. Statist.* **11** 186-192

Hotelling, H. (1931). The generalization of "Student's" ratio. *Ann. Math. Statist.* **2** 360-378.

Neyman, J. and Pearson, E.S. (1928). The use and interpretation of certain test criteria for purposes of statistical inference, Part I. *Biometrika* **20a** 175-240.

Stein, C. (1945). A two-sample test for a linear hypothesis whose power is independent of the variance. *Ann. Math. Statist.* **16** 243-258.

Previous Tables, Charts and Programs

Asaki, Z, Kondo, Y., and Narita, T. (1970). On the power of F-test in analysis of variance-I; Power of F-test on the main effects. *Reports of Statist. Appl. Res., Union of Japanese Scientists and Engineers* **17** 57-66.

Asaki, Z. and Kondo Y. (1971). On the power of the F-test in analysis of variance-II; Power of F-test on the effect of two factor interactions. *Reports of Statist. Appl. Res., Union of Japanese Scientists and Engineers* **18** 1-8.

Bargmann, R.E. and Ghosh, S.P. (1963). Statistical distribution programs for a computer language. *I.B.M. Watson Res. Center* Res. Rep. RC-1094.

Bargmann, R.E. and Ghosh, S.P. (1964). Noncentral statistical distribution programs for a computer language. *I.B.M. Watson Res. Center* Res. Rep. RC-1231.

Bargmann, R.E. and Thomas, C.G. (1971). Comparison of noncentral distribution programs. *Themis Tech. Report No. 10*, U. Georgia.

Bouver, H. and Bargmann, R.E. (1975). Computational algorithms for the evaluation of statistical distribution functions. *Themis Tech. Report No. 36*, U. Georgia.

Bowman, K.O. and Kastenbaum, M.A. (1975). Sample size requirement: single and double classification experiments. *Selected Tables in Mathematical Statistics, Vol. 3*, Harter, H.L. and Owen, D.B., eds. (Providence: American Mathematical Society). 111-232.

Bratcher, T.L., Moran, M.A., and Zimmer, W.I. (1970). Tables of sample sizes in the analysis of variance. *J. Qual. Tech.* **2** 156-164.

Dasgupta, P. (1968). Tables of the noncentrality parameter of F-test as a function of power. *Sankhyā, Series B* **30** 73-82.

Duncan, A.J. (1957). Charts of the 10% and 50% points of the operating characteristic curves for fixed effects analysis of variance F-tests, $\alpha = 0.01$ and .05. *J. Amer. Statist. Assoc.* **52** 345-349.

Feldt, L.S. and Mahmoud, M.W. (1958). Power function charts for specifying numbers of observations in analysis of variance for fixed effects. *Ann. Math. Statist.* **29** 871-877.

Fox, M. (1956). Charts of the power of the F-test. *Ann. Math. Statist.* **29** 484-497.

Ifram, A.F. (1971). Tables of the noncentral F and beta variables. *Jordon Res. Council* **1** 181-207.

Kastenbaum, M.A., Hoel, D.G., and Bowman, K.O. (1970a). Sample size requirement: one-way analysis of variance. *Biometrika* **57** 421-430.

Kastenbaum, M.A., Hoel, D.G., and Bowman, K.O. (1970b). Sample size requirement: randomized block designs. *Biometrika* **57** 573-577.

Keuls, M. (1960a). Tabellen in nomogrammen voor het orderscheidingsvermogen van de 5% en 10%-F-toets voor het gebruik bij de gewarde blokke proef. *Statistica Neerlandica* **14** 127-150 (Dutch with English summary).

Keuls M. (1960b). La puissance du critère F dans l'analyse de la variance de plans en blocs an hasard. Nomogrammes pur le choix du nombre de répétitions. *Biometrie Praximetrie* **1** 65-80 (Dutch and English summaries).

Keuls M. (1960c). La puissance du critère F dans l'analyse de la variance de plans en blocs an hasard. Nomogrammes pur le choix du nombre de répétitions. *Bull. de l'Inst. Agron. et des Stations de Recherches de Gembloux* **1** 256-271 (English and German summaries).

Lachenbruch, P.A. (1967). The non-central F-distribution: Some extensions of Tang's tables. *Dept. of Biostatistics, U. of North Carolina* Mimeo Series No. 531.

Lehmer, E. (1944). Inverse table of probabilities of errors of the second kind. *Ann. of Math. Statist.* **15** 388-398.

Owen, D.B. (1962). *Handbook of Statistical Tables.* (Reading, Mass: Addison-Wesley Publishing Co., Inc.).

Pearson, E.S. and Hartley, H.O. (1951). Charts of the power function for analysis of variance tests, derived from the non-central F-distribution. *Biometrika* **38** 112-130.

Pearson, E.S. and Hartley, H.O. (1972). *Biometrika Tables for Statisticians, Vol. 2.* (London: Cambridge University Press).

Rodger, R.S. (1974). Multiple contrasts, factors, error rates and power. *Br. J. math. statist. Psychol.* **27** 179-198.

Rodger, R.S. (1975a). The number of non-zero, *post hoc* contrasts from ANOVA and error-rate I. *Br. J. math. statist. Psychol.* **28** 71-78.

Rodger, R.S. (1975b). Setting rejection rate for contrasts selected *post hoc* when some nulls are false. *Br. J. math. statist. Psychol.* **28** 214-232.

Rodger, R.S. (1976). Tables of Stein's non-central parameter $D\beta$; ν_1, ν_2 required to set power for numerical alternatives to H_0 tested by two-stage sampling anova. *J. statist. Comput. Simul.* **5** 1-22.

Rodger, R.S. (1978). Two-stage sampling to set sample size for *post hoc* tests in ANOVA with decision-based error rates. *Br. J. math. statist. Psychol.* **31** 153-178.

Scheffé, H. (1959). *The Analysis of Variance.* (New York: John Wiley & Sons, Inc.).

Tang, P.C. (1938). The power function of the analysis of variance test with tables and illustrations of their use. *Stat. Res. Mem.* **2** 126-249 and tables.

Tiku, M.L. (1967). Tables of the power of the F-test. *J. Amer. Statist. Assoc.* **62** 525-539.

Tiku, M.L. (1972). More tables of the power of the F-test. *J. Amer. Statist. Assoc.* **67** 709-710.

Computational Methods

Abramowitz, M. and Stegun, I.A. (editors) (1968). *Handbook of Mathematical Functions* (seventh printing). AMS 55, U.S. Department of Commerce, National Bureau of Standards, Washington, D.C.

Beasley, J.D. and Springer, S.G. (1985). The percentage points of the normal distribution, Algorithm AS 111. *Applied Statistics Algorithms*, Griffiths P. and Hill, I.D., eds. (Chichester: Ellis Horwood Limited). 188-191.

Best, D.J. and Roberts, D.E. (1985). The percentage points of the χ^2 distribution, Algorithm AS 91. *Applied Statistics Algorithms*, Griffiths P. and Hill, I.D., eds. (Chichester: Ellis Horwood Limited). 157-161.

Bhattacharjee, C.P. (1970). The incomplete gamma integral. Algorithm AS 32. *Appl. Statist.* **19** 285-287.

Goldstein, R.B. (1973). Chi-square quantiles, Algorithm 451. *Comm. ACM* **16** 482-484.

Hill, G.W. (1970). Student's t-quantiles, Algorithm 396. *Comm. ACM* **13** 619-620.

Lau, C-L (1980). A simple series for the incomplete gamma integral, Algorithm AS 147. *Appl. Statist.* **29** 113-114.

Majumder, K.L. and Bhattacharjee, G.P. (1985a). The incomplete beta integral, Algorithm AS 63. *Applied Statistics Algorithms*, Griffiths P. and Hill, I.D., eds. (Chichester: Ellis Horwood Limited). 117-120.

Majumder, K.L. and Bhattacharjee, G.P. (1985b). Inverse of the incomplete beta function ratio, Algorithm AS 64/AS 109. *Applied Statistics Algorithms*, Griffiths P. and Hill, I.D., eds. (Chichester: Ellis Horwood Limited). 121-125.

McCracken, D.D. and Dorn, W.S. (1964). *Numerical Methods and Fortran Programming* (New York: John Wiley & Sons, Inc.).

Narula, S.C. and Desu, M.M. (1981). Computation of probability and non-centrality parameter of a non-central chi-square distribution, Algorithm AS 170. *Appl. Statist.* **30** 349-352.

Narula, S.C. and Weistroffer, H.R.(1986). Computation of probability and non-centrality parameter of a non-central F-distribution. *Commun. Statist.-Simula.* **15** 871-878.

Norton, V. (1983). A simple algorithm for computing the non-central F-distribution. *Appl. Statist.*, **32** 84-85.

Odeh, R.E. and Evans, O.J. (1972). Some rational approximations to the upper percentage points of the normal distribution. *Proc. Second Manitoba Conference on Numerical Mathematics,* University of Winnipeg, Manitoba 311-318.

Odeh R.E. and Evans, O.J. (1974), The percentage points of the normal distribution. Algorithm AS 70. *Appl. Statist.* **23** 96-97.

Tiku, M.L. (1965). Laguerre series forms of non-central χ^2 and F distributions. *Biometrika* **52** 415-427.

Tiku, M.L. (1966). A note on approximating to the non-central F distribution. *Biometrika* **53** 606-610.

PART TWO

TABLE 1.1

$\alpha = 0.50$ CRITICAL VALUES OF THE F-DISTRIBUTION

	f_1											
$f_2 \downarrow$	1	2	3	4	5	6	7	8	9	10	11	12
6	.5149	.7798	.8858	.9419	.9765	1.000	1.017	1.030	1.040	1.048	1.054	1.060
7	.5057	.7665	.8709	.9262	.9603	.9833	1.000	1.013	1.022	1.030	1.037	1.042
8	.4990	.7568	.8600	.9146	.9483	.9711	.9876	1.000	1.010	1.018	1.024	1.029
9	.4938	.7494	.8517	.9058	.9392	.9617	.9781	.9904	1.000	1.008	1.014	1.019
10	.4897	.7435	.8451	.8988	.9319	.9544	.9705	.9828	.9923	1.000	1.006	1.012
12	.4837	.7348	.8353	.8885	.9212	.9434	.9594	.9715	.9810	.9886	.9948	1.000
14	.4794	.7286	.8284	.8812	.9137	.9357	.9516	.9636	.9730	.9805	.9867	.9919
16	.4763	.7241	.8233	.8758	.9081	.9300	.9458	.9577	.9670	.9745	.9807	.9858
18	.4738	.7205	.8194	.8716	.9038	.9256	.9413	.9532	.9625	.9699	.9760	.9812
20	.4719	.7177	.8162	.8683	.9004	.9221	.9378	.9496	.9588	.9663	.9724	.9775
22	.4703	.7155	.8137	.8656	.8976	.9192	.9349	.9467	.9559	.9633	.9694	.9744
24	.4690	.7136	.8115	.8633	.8953	.9169	.9325	.9442	.9534	.9608	.9669	.9719
26	.4679	.7120	.8097	.8615	.8933	.9149	.9304	.9422	.9513	.9587	.9648	.9698
28	.4670	.7106	.8082	.8598	.8916	.9132	.9287	.9404	.9496	.9569	.9630	.9680
30	.4662	.7094	.8069	.8584	.8902	.9117	.9272	.9389	.9480	.9554	.9614	.9665
35	.4645	.7071	.8042	.8556	.8873	.9087	.9242	.9359	.9450	.9523	.9583	.9634
40	.4633	.7053	.8023	.8536	.8852	.9065	.9220	.9336	.9427	.9500	.9560	.9610
50	.4616	.7028	.7995	.8507	.8822	.9035	.9189	.9305	.9395	.9468	.9528	.9578
60	.4605	.7012	.7977	.8487	.8802	.9014	.9168	.9284	.9374	.9447	.9507	.9557
70	.4597	.7001	.7964	.8474	.8787	.9000	.9153	.9269	.9359	.9432	.9492	.9541
80	.4591	.6992	.7954	.8463	.8777	.8989	.9142	.9258	.9348	.9421	.9480	.9530
90	.4586	.6985	.7947	.8455	.8769	.8981	.9134	.9249	.9339	.9412	.9471	.9521
120	.4577	.6972	.7932	.8439	.8752	.8964	.9116	.9232	.9322	.9394	.9454	.9503
240	.4563	.6952	.7909	.8415	.8727	.8939	.9091	.9206	.9296	.9368	.9427	.9477
∞	.4549	.6931	.7887	.8392	.8703	.8914	.9065	.9180	.9270	.9342	.9401	.9450

	f_1											
$f_2 \downarrow$	14	16	18	20	22	24	30	40	60	90	120	∞
6	1.069	1.075	1.080	1.084	1.088	1.091	1.097	1.103	1.109	1.114	1.116	1.122
7	1.051	1.057	1.062	1.066	1.070	1.072	1.079	1.085	1.091	1.095	1.097	1.103
8	1.038	1.044	1.049	1.053	1.056	1.059	1.065	1.071	1.077	1.081	1.083	1.089
9	1.028	1.034	1.039	1.043	1.046	1.049	1.055	1.061	1.067	1.071	1.073	1.079
10	1.020	1.026	1.031	1.035	1.038	1.041	1.047	1.053	1.059	1.062	1.064	1.070
12	1.008	1.014	1.019	1.023	1.026	1.029	1.035	1.041	1.046	1.050	1.052	1.058
14	1.000	1.006	1.011	1.015	1.018	1.020	1.026	1.032	1.038	1.042	1.044	1.050
16	.9939	1.000	1.005	1.009	1.012	1.014	1.020	1.026	1.032	1.035	1.037	1.043
18	.9892	.9953	1.000	1.004	1.007	1.009	1.015	1.021	1.027	1.030	1.032	1.038
20	.9855	.9915	.9962	1.000	1.003	1.006	1.011	1.017	1.023	1.027	1.029	1.034
22	.9824	.9885	.9932	.9969	1.000	1.003	1.008	1.014	1.020	1.023	1.025	1.031
24	.9799	.9859	.9906	.9944	.9974	1.000	1.006	1.011	1.017	1.021	1.023	1.028
26	.9778	.9838	.9885	.9922	.9953	.9978	1.003	1.009	1.015	1.019	1.020	1.026
28	.9760	.9820	.9866	.9904	.9934	.9960	1.002	1.007	1.013	1.017	1.019	1.024
30	.9744	.9804	.9850	.9888	.9918	.9944	1.000	1.006	1.011	1.015	1.017	1.023
35	.9713	.9772	.9819	.9856	.9886	.9912	.9968	1.002	1.008	1.012	1.014	1.019
40	.9689	.9749	.9795	.9832	.9863	.9888	.9944	1.000	1.006	1.009	1.011	1.017
50	.9657	.9716	.9762	.9799	.9830	.9855	.9911	.9966	1.002	1.006	1.008	1.013
60	.9635	.9694	.9740	.9777	.9808	.9833	.9888	.9944	1.000	1.004	1.006	1.011
70	.9620	.9679	.9725	.9762	.9792	.9817	.9873	.9928	.9984	1.002	1.004	1.010
80	.9608	.9667	.9713	.9750	.9780	.9805	.9861	.9916	.9972	1.001	1.003	1.008
90	.9599	.9658	.9704	.9741	.9771	.9796	.9852	.9907	.9963	1.000	1.002	1.007
120	.9581	.9640	.9686	.9723	.9753	.9778	.9833	.9889	.9944	.9981	1.000	1.006
240	.9555	.9613	.9659	.9696	.9726	.9751	.9806	.9861	.9917	.9954	.9972	1.003
∞	.9528	.9587	.9632	.9669	.9699	.9724	.9779	.9834	.9889	.9926	.9944	1.000

TABLE 1.2

$\alpha = 0.25$ CRITICAL VALUES OF THE F-DISTRIBUTION

	f_1											
$f_2 \downarrow$	1	2	3	4	5	6	7	8	9	10	11	12
6	1.621	1.762	1.784	1.787	1.785	1.782	1.779	1.776	1.773	1.771	1.769	1.767
7	1.573	1.701	1.717	1.716	1.711	1.706	1.701	1.697	1.693	1.690	1.687	1.684
8	1.538	1.657	1.668	1.664	1.658	1.651	1.645	1.640	1.635	1.631	1.627	1.624
9	1.512	1.624	1.632	1.625	1.617	1.609	1.602	1.596	1.591	1.586	1.582	1.579
10	1.491	1.598	1.603	1.595	1.585	1.576	1.569	1.562	1.556	1.551	1.547	1.543
12	1.461	1.560	1.561	1.550	1.539	1.529	1.520	1.512	1.505	1.500	1.495	1.490
14	1.440	1.533	1.532	1.519	1.507	1.495	1.485	1.477	1.470	1.463	1.458	1.453
16	1.425	1.514	1.510	1.497	1.483	1.471	1.460	1.451	1.443	1.437	1.431	1.426
18	1.413	1.499	1.494	1.479	1.464	1.452	1.441	1.431	1.423	1.416	1.410	1.404
20	1.404	1.487	1.481	1.465	1.450	1.437	1.425	1.415	1.407	1.399	1.393	1.387
22	1.396	1.477	1.470	1.454	1.438	1.424	1.413	1.402	1.394	1.386	1.379	1.374
24	1.390	1.470	1.462	1.445	1.428	1.414	1.402	1.392	1.383	1.375	1.368	1.362
26	1.384	1.463	1.454	1.437	1.420	1.406	1.393	1.383	1.374	1.366	1.359	1.352
28	1.380	1.457	1.448	1.430	1.413	1.399	1.386	1.375	1.366	1.358	1.350	1.344
30	1.376	1.452	1.443	1.424	1.407	1.392	1.380	1.369	1.359	1.351	1.343	1.337
35	1.368	1.443	1.432	1.413	1.395	1.380	1.367	1.355	1.345	1.337	1.329	1.323
40	1.363	1.435	1.424	1.404	1.386	1.371	1.357	1.345	1.335	1.327	1.319	1.312
50	1.355	1.425	1.413	1.393	1.374	1.358	1.344	1.332	1.321	1.312	1.304	1.297
60	1.349	1.419	1.405	1.385	1.366	1.349	1.335	1.323	1.312	1.303	1.294	1.287
70	1.346	1.414	1.400	1.379	1.360	1.343	1.329	1.316	1.305	1.296	1.287	1.280
80	1.343	1.411	1.396	1.375	1.355	1.338	1.324	1.311	1.300	1.291	1.282	1.275
90	1.341	1.408	1.393	1.372	1.352	1.335	1.320	1.307	1.296	1.287	1.278	1.270
120	1.336	1.402	1.387	1.365	1.345	1.328	1.313	1.300	1.289	1.279	1.270	1.262
240	1.330	1.394	1.378	1.356	1.335	1.317	1.302	1.289	1.277	1.267	1.258	1.250
∞	1.323	1.386	1.369	1.346	1.325	1.307	1.291	1.277	1.265	1.255	1.246	1.237

	f_1											
$f_2 \downarrow$	14	16	18	20	22	24	30	40	60	90	120	∞
6	1.764	1.761	1.759	1.757	1.755	1.754	1.751	1.748	1.744	1.742	1.741	1.737
7	1.680	1.676	1.674	1.671	1.669	1.667	1.663	1.659	1.655	1.652	1.650	1.645
8	1.619	1.615	1.612	1.609	1.606	1.604	1.600	1.595	1.589	1.586	1.584	1.578
9	1.573	1.568	1.564	1.561	1.558	1.556	1.551	1.545	1.539	1.535	1.533	1.526
10	1.537	1.531	1.527	1.523	1.520	1.518	1.512	1.506	1.499	1.494	1.492	1.484
12	1.483	1.477	1.472	1.468	1.464	1.461	1.454	1.447	1.439	1.434	1.431	1.422
14	1.445	1.438	1.433	1.428	1.425	1.421	1.414	1.405	1.397	1.391	1.387	1.377
16	1.417	1.410	1.404	1.399	1.395	1.391	1.383	1.374	1.365	1.358	1.354	1.343
18	1.395	1.388	1.381	1.376	1.372	1.368	1.359	1.350	1.340	1.332	1.328	1.316
20	1.378	1.370	1.363	1.358	1.353	1.349	1.340	1.330	1.319	1.311	1.307	1.294
22	1.364	1.355	1.349	1.343	1.338	1.334	1.324	1.314	1.303	1.294	1.290	1.276
24	1.352	1.343	1.337	1.331	1.326	1.321	1.311	1.300	1.289	1.280	1.275	1.261
26	1.342	1.333	1.326	1.320	1.315	1.311	1.300	1.289	1.277	1.268	1.263	1.247
28	1.333	1.325	1.317	1.311	1.306	1.301	1.291	1.279	1.266	1.257	1.252	1.236
30	1.326	1.317	1.310	1.303	1.298	1.293	1.282	1.270	1.257	1.247	1.242	1.226
35	1.311	1.302	1.294	1.288	1.282	1.277	1.266	1.253	1.239	1.228	1.223	1.205
40	1.300	1.291	1.283	1.276	1.270	1.265	1.253	1.240	1.225	1.214	1.208	1.188
50	1.285	1.275	1.266	1.259	1.253	1.248	1.235	1.221	1.205	1.193	1.186	1.164
60	1.274	1.264	1.255	1.248	1.242	1.236	1.223	1.208	1.191	1.178	1.172	1.147
70	1.267	1.257	1.248	1.240	1.234	1.228	1.214	1.199	1.181	1.168	1.161	1.135
80	1.262	1.251	1.242	1.234	1.227	1.222	1.208	1.192	1.174	1.160	1.152	1.124
90	1.257	1.246	1.237	1.229	1.223	1.217	1.202	1.186	1.168	1.153	1.145	1.116
120	1.249	1.237	1.228	1.220	1.213	1.207	1.192	1.175	1.156	1.140	1.131	1.099
240	1.236	1.224	1.214	1.206	1.198	1.192	1.176	1.158	1.137	1.119	1.109	1.067
∞	1.223	1.211	1.200	1.191	1.184	1.177	1.160	1.140	1.116	1.096	1.084	1.000

TABLE 1.3
$\alpha = 0.10$ CRITICAL VALUES OF THE F-DISTRIBUTION

	f_1											
$f_2 \downarrow$	1	2	3	4	5	6	7	8	9	10	11	12
6	3.776	3.463	3.289	3.181	3.108	3.055	3.014	2.983	2.958	2.937	2.920	2.905
7	3.589	3.257	3.074	2.961	2.883	2.827	2.785	2.752	2.725	2.703	2.684	2.668
8	3.458	3.113	2.924	2.806	2.726	2.668	2.624	2.589	2.561	2.538	2.519	2.502
9	3.360	3.006	2.813	2.693	2.611	2.551	2.505	2.469	2.440	2.416	2.396	2.379
10	3.285	2.924	2.728	2.605	2.522	2.461	2.414	2.377	2.347	2.323	2.302	2.284
12	3.177	2.807	2.606	2.480	2.394	2.331	2.283	2.245	2.214	2.188	2.166	2.147
14	3.102	2.726	2.522	2.395	2.307	2.243	2.193	2.154	2.122	2.095	2.073	2.054
16	3.048	2.668	2.462	2.333	2.244	2.178	2.128	2.088	2.055	2.028	2.005	1.985
18	3.007	2.624	2.416	2.286	2.196	2.130	2.079	2.038	2.005	1.977	1.954	1.933
20	2.975	2.589	2.380	2.249	2.158	2.091	2.040	1.999	1.965	1.937	1.913	1.892
22	2.949	2.561	2.351	2.219	2.128	2.060	2.008	1.967	1.933	1.904	1.880	1.859
24	2.927	2.538	2.327	2.195	2.103	2.035	1.983	1.941	1.906	1.877	1.853	1.832
26	2.909	2.519	2.307	2.174	2.082	2.014	1.961	1.919	1.884	1.855	1.830	1.809
28	2.894	2.503	2.291	2.157	2.064	1.996	1.943	1.900	1.865	1.836	1.811	1.790
30	2.881	2.489	2.276	2.142	2.049	1.980	1.927	1.884	1.849	1.819	1.794	1.773
35	2.855	2.461	2.247	2.113	2.019	1.950	1.896	1.852	1.817	1.787	1.761	1.739
40	2.835	2.440	2.226	2.091	1.997	1.927	1.873	1.829	1.793	1.763	1.737	1.715
50	2.809	2.412	2.197	2.061	1.966	1.895	1.840	1.796	1.760	1.729	1.703	1.680
60	2.791	2.393	2.177	2.041	1.946	1.875	1.819	1.775	1.738	1.707	1.680	1.657
70	2.779	2.380	2.164	2.027	1.931	1.860	1.804	1.760	1.723	1.691	1.665	1.641
80	2.769	2.370	2.154	2.016	1.921	1.849	1.793	1.748	1.711	1.680	1.653	1.629
90	2.762	2.363	2.146	2.008	1.912	1.841	1.785	1.739	1.702	1.670	1.643	1.620
120	2.748	2.347	2.130	1.992	1.896	1.824	1.767	1.722	1.684	1.652	1.625	1.601
240	2.727	2.325	2.107	1.968	1.871	1.799	1.742	1.696	1.658	1.625	1.598	1.573
∞	2.706	2.303	2.084	1.945	1.847	1.774	1.717	1.670	1.632	1.599	1.570	1.546

	f_1											
$f_2 \downarrow$	14	16	18	20	22	24	30	40	60	90	120	∞
6	2.881	2.863	2.848	2.836	2.827	2.818	2.800	2.781	2.762	2.749	2.742	2.722
7	2.643	2.623	2.607	2.595	2.584	2.575	2.555	2.535	2.514	2.500	2.493	2.471
8	2.475	2.455	2.438	2.425	2.413	2.404	2.383	2.361	2.339	2.324	2.316	2.293
9	2.351	2.329	2.312	2.298	2.287	2.277	2.255	2.232	2.208	2.192	2.184	2.159
10	2.255	2.233	2.215	2.201	2.189	2.178	2.155	2.132	2.107	2.090	2.082	2.055
12	2.117	2.094	2.075	2.060	2.047	2.036	2.011	1.986	1.960	1.942	1.932	1.904
14	2.022	1.998	1.978	1.962	1.949	1.938	1.912	1.885	1.857	1.838	1.828	1.797
16	1.953	1.928	1.908	1.891	1.877	1.866	1.839	1.811	1.782	1.761	1.751	1.718
18	1.900	1.875	1.854	1.837	1.823	1.810	1.783	1.754	1.723	1.702	1.691	1.657
20	1.859	1.833	1.811	1.794	1.779	1.767	1.738	1.708	1.677	1.655	1.643	1.607
22	1.825	1.798	1.777	1.759	1.744	1.731	1.702	1.671	1.639	1.616	1.604	1.567
24	1.797	1.770	1.748	1.730	1.715	1.702	1.672	1.641	1.607	1.584	1.571	1.533
26	1.774	1.747	1.724	1.706	1.690	1.677	1.647	1.615	1.581	1.556	1.544	1.504
28	1.754	1.726	1.704	1.685	1.669	1.656	1.625	1.592	1.558	1.533	1.520	1.478
30	1.737	1.709	1.686	1.667	1.651	1.638	1.606	1.573	1.538	1.512	1.499	1.456
35	1.703	1.674	1.651	1.632	1.615	1.601	1.569	1.535	1.497	1.471	1.457	1.411
40	1.678	1.649	1.625	1.605	1.588	1.574	1.541	1.506	1.467	1.439	1.425	1.377
50	1.643	1.613	1.588	1.568	1.551	1.536	1.502	1.465	1.424	1.395	1.379	1.327
60	1.619	1.589	1.564	1.543	1.526	1.511	1.476	1.437	1.395	1.364	1.348	1.291
70	1.603	1.572	1.547	1.526	1.508	1.493	1.457	1.418	1.374	1.342	1.325	1.265
80	1.590	1.559	1.534	1.513	1.495	1.479	1.443	1.403	1.358	1.325	1.307	1.245
90	1.581	1.550	1.524	1.503	1.484	1.468	1.432	1.391	1.346	1.312	1.293	1.228
120	1.562	1.530	1.504	1.482	1.463	1.447	1.409	1.368	1.320	1.284	1.265	1.193
240	1.533	1.501	1.474	1.451	1.432	1.415	1.376	1.332	1.281	1.242	1.219	1.130
∞	1.505	1.471	1.444	1.421	1.401	1.383	1.342	1.295	1.240	1.195	1.169	1.000

TABLE 1.4

$\alpha = 0.05$ CRITICAL VALUES OF THE F-DISTRIBUTION

	f_1											
$f_2 \downarrow$	1	2	3	4	5	6	7	8	9	10	11	12
6	5.987	5.143	4.757	4.534	4.387	4.284	4.207	4.147	4.099	4.060	4.027	4.000
7	5.591	4.737	4.347	4.120	3.972	3.866	3.787	3.726	3.677	3.637	3.603	3.575
8	5.318	4.459	4.066	3.838	3.687	3.581	3.500	3.438	3.388	3.347	3.313	3.284
9	5.117	4.256	3.863	3.633	3.482	3.374	3.293	3.230	3.179	3.137	3.102	3.073
10	4.965	4.103	3.708	3.478	3.326	3.217	3.135	3.072	3.020	2.978	2.943	2.913
12	4.747	3.885	3.490	3.259	3.106	2.996	2.913	2.849	2.796	2.753	2.717	2.687
14	4.600	3.739	3.344	3.112	2.958	2.848	2.764	2.699	2.646	2.602	2.565	2.534
16	4.494	3.634	3.239	3.007	2.852	2.741	2.657	2.591	2.538	2.494	2.456	2.425
18	4.414	3.555	3.160	2.928	2.773	2.661	2.577	2.510	2.456	2.412	2.374	2.342
20	4.351	3.493	3.098	2.866	2.711	2.599	2.514	2.447	2.393	2.348	2.310	2.278
22	4.301	3.443	3.049	2.817	2.661	2.549	2.464	2.397	2.342	2.297	2.259	2.226
24	4.260	3.403	3.009	2.776	2.621	2.508	2.423	2.355	2.300	2.255	2.216	2.183
26	4.225	3.369	2.975	2.743	2.587	2.474	2.388	2.321	2.265	2.220	2.181	2.148
28	4.196	3.340	2.947	2.714	2.558	2.445	2.359	2.291	2.236	2.190	2.151	2.118
30	4.171	3.316	2.922	2.690	2.534	2.421	2.334	2.266	2.211	2.165	2.126	2.092
35	4.121	3.267	2.874	2.641	2.485	2.372	2.285	2.217	2.161	2.114	2.075	2.041
40	4.085	3.232	2.839	2.606	2.449	2.336	2.249	2.180	2.124	2.077	2.038	2.003
50	4.034	3.183	2.790	2.557	2.400	2.286	2.199	2.130	2.073	2.026	1.986	1.952
60	4.001	3.150	2.758	2.525	2.368	2.254	2.167	2.097	2.040	1.993	1.952	1.917
70	3.978	3.128	2.736	2.503	2.346	2.231	2.143	2.074	2.017	1.969	1.928	1.893
80	3.960	3.111	2.719	2.486	2.329	2.214	2.126	2.056	1.999	1.951	1.910	1.875
90	3.947	3.098	2.706	2.473	2.316	2.201	2.113	2.043	1.986	1.938	1.897	1.861
120	3.920	3.072	2.680	2.447	2.290	2.175	2.087	2.016	1.959	1.910	1.869	1.834
240	3.880	3.033	2.642	2.409	2.252	2.136	2.048	1.977	1.919	1.870	1.829	1.793
∞	3.841	2.996	2.605	2.372	2.214	2.099	2.010	1.938	1.880	1.831	1.789	1.752

	f_1											
$f_2 \downarrow$	14	16	18	20	22	24	30	40	60	90	120	∞
6	3.956	3.922	3.896	3.874	3.856	3.841	3.808	3.774	3.740	3.716	3.705	3.669
7	3.529	3.494	3.467	3.445	3.426	3.410	3.376	3.340	3.304	3.280	3.267	3.230
8	3.237	3.202	3.173	3.150	3.131	3.115	3.079	3.043	3.005	2.980	2.967	2.928
9	3.025	2.989	2.960	2.936	2.917	2.900	2.864	2.826	2.787	2.761	2.748	2.707
10	2.865	2.828	2.798	2.774	2.754	2.737	2.700	2.661	2.621	2.594	2.580	2.538
12	2.637	2.599	2.568	2.544	2.523	2.505	2.466	2.426	2.384	2.356	2.341	2.296
14	2.484	2.445	2.413	2.388	2.367	2.349	2.308	2.266	2.223	2.193	2.178	2.131
16	2.373	2.333	2.302	2.276	2.254	2.235	2.194	2.151	2.106	2.075	2.059	2.010
18	2.290	2.250	2.217	2.191	2.168	2.150	2.107	2.063	2.017	1.985	1.968	1.917
20	2.225	2.184	2.151	2.124	2.102	2.082	2.039	1.994	1.946	1.913	1.896	1.843
22	2.173	2.131	2.098	2.071	2.048	2.028	1.984	1.938	1.889	1.856	1.838	1.783
24	2.130	2.088	2.054	2.027	2.003	1.984	1.939	1.892	1.842	1.808	1.790	1.733
26	2.094	2.052	2.018	1.990	1.966	1.946	1.901	1.853	1.803	1.767	1.749	1.691
28	2.064	2.021	1.987	1.959	1.935	1.915	1.869	1.820	1.769	1.733	1.714	1.654
30	2.037	1.995	1.960	1.932	1.908	1.887	1.841	1.792	1.740	1.703	1.683	1.622
35	1.986	1.942	1.907	1.878	1.854	1.833	1.786	1.735	1.681	1.643	1.623	1.558
40	1.948	1.904	1.868	1.839	1.814	1.793	1.744	1.693	1.637	1.597	1.577	1.509
50	1.895	1.850	1.814	1.784	1.759	1.737	1.687	1.634	1.576	1.534	1.511	1.438
60	1.860	1.815	1.778	1.748	1.722	1.700	1.649	1.594	1.534	1.491	1.467	1.389
70	1.836	1.790	1.753	1.722	1.696	1.674	1.622	1.566	1.505	1.459	1.435	1.353
80	1.817	1.772	1.734	1.703	1.677	1.654	1.602	1.545	1.482	1.436	1.411	1.325
90	1.803	1.757	1.720	1.688	1.662	1.639	1.586	1.528	1.465	1.417	1.391	1.302
120	1.775	1.728	1.690	1.659	1.632	1.608	1.554	1.495	1.429	1.379	1.352	1.254
240	1.733	1.686	1.647	1.614	1.587	1.563	1.507	1.445	1.375	1.320	1.290	1.170
∞	1.692	1.644	1.604	1.571	1.542	1.517	1.459	1.394	1.318	1.257	1.221	1.000

TABLE 1.5

$\alpha = 0.025$ CRITICAL VALUES OF THE F-DISTRIBUTION

	f_1											
$f_2 \downarrow$	1	2	3	4	5	6	7	8	9	10	11	12
6	8.813	7.260	6.599	6.227	5.988	5.820	5.695	5.600	5.523	5.461	5.410	5.366
7	8.073	6.542	5.890	5.523	5.285	5.119	4.995	4.899	4.823	4.761	4.709	4.666
8	7.571	6.059	5.416	5.053	4.817	4.652	4.529	4.433	4.357	4.295	4.243	4.200
9	7.209	5.715	5.078	4.718	4.484	4.320	4.197	4.102	4.026	3.964	3.912	3.868
10	6.937	5.456	4.826	4.468	4.236	4.072	3.950	3.855	3.779	3.717	3.665	3.621
12	6.554	5.096	4.474	4.121	3.891	3.728	3.607	3.512	3.436	3.374	3.321	3.277
14	6.298	4.857	4.242	3.892	3.663	3.501	3.380	3.285	3.209	3.147	3.095	3.050
16	6.115	4.687	4.077	3.729	3.502	3.341	3.219	3.125	3.049	2.986	2.934	2.889
18	5.978	4.560	3.954	3.608	3.382	3.221	3.100	3.005	2.929	2.866	2.814	2.769
20	5.871	4.461	3.859	3.515	3.289	3.128	3.007	2.913	2.837	2.774	2.721	2.676
22	5.786	4.383	3.783	3.440	3.215	3.055	2.934	2.839	2.763	2.700	2.647	2.602
24	5.717	4.319	3.721	3.379	3.155	2.995	2.874	2.779	2.703	2.640	2.586	2.541
26	5.659	4.265	3.670	3.329	3.105	2.945	2.824	2.729	2.653	2.590	2.536	2.491
28	5.610	4.221	3.626	3.286	3.063	2.903	2.782	2.687	2.611	2.547	2.494	2.448
30	5.568	4.182	3.589	3.250	3.026	2.867	2.746	2.651	2.575	2.511	2.458	2.412
35	5.485	4.106	3.517	3.179	2.956	2.796	2.676	2.581	2.504	2.440	2.387	2.341
40	5.424	4.051	3.463	3.126	2.904	2.744	2.624	2.529	2.452	2.388	2.334	2.288
50	5.340	3.975	3.390	3.054	2.833	2.674	2.553	2.458	2.381	2.317	2.263	2.216
60	5.286	3.925	3.343	3.008	2.786	2.627	2.507	2.412	2.334	2.270	2.216	2.169
70	5.247	3.890	3.309	2.975	2.754	2.595	2.474	2.379	2.302	2.237	2.183	2.136
80	5.218	3.864	3.284	2.950	2.730	2.571	2.450	2.355	2.277	2.213	2.158	2.111
90	5.196	3.844	3.265	2.932	2.711	2.552	2.432	2.336	2.259	2.194	2.140	2.092
120	5.152	3.805	3.227	2.894	2.674	2.515	2.395	2.299	2.222	2.157	2.102	2.055
240	5.088	3.746	3.171	2.839	2.620	2.461	2.341	2.245	2.167	2.102	2.047	1.999
∞	5.024	3.689	3.116	2.786	2.567	2.408	2.288	2.192	2.114	2.048	1.993	1.945

	f_1											
$f_2 \downarrow$	14	16	18	20	22	24	30	40	60	90	120	∞
6	5.297	5.244	5.202	5.168	5.141	5.117	5.065	5.012	4.959	4.923	4.904	4.849
7	4.596	4.543	4.501	4.467	4.439	4.415	4.362	4.309	4.254	4.218	4.199	4.142
8	4.130	4.076	4.034	3.999	3.971	3.947	3.894	3.840	3.784	3.747	3.728	3.670
9	3.798	3.744	3.701	3.667	3.638	3.614	3.560	3.505	3.449	3.411	3.392	3.333
10	3.550	3.496	3.453	3.419	3.390	3.365	3.311	3.255	3.198	3.160	3.140	3.080
12	3.206	3.152	3.108	3.073	3.043	3.019	2.963	2.906	2.848	2.808	2.787	2.725
14	2.979	2.923	2.879	2.844	2.814	2.789	2.732	2.674	2.614	2.573	2.552	2.487
16	2.817	2.761	2.717	2.681	2.651	2.625	2.568	2.509	2.447	2.405	2.383	2.316
18	2.696	2.640	2.596	2.559	2.529	2.503	2.445	2.384	2.321	2.278	2.256	2.187
20	2.603	2.547	2.501	2.464	2.434	2.408	2.349	2.287	2.223	2.179	2.156	2.085
22	2.528	2.472	2.426	2.389	2.358	2.331	2.272	2.210	2.145	2.099	2.076	2.003
24	2.468	2.411	2.365	2.327	2.296	2.269	2.209	2.146	2.080	2.034	2.010	1.935
26	2.417	2.360	2.314	2.276	2.244	2.217	2.157	2.093	2.026	1.979	1.954	1.878
28	2.374	2.317	2.270	2.232	2.201	2.174	2.112	2.048	1.980	1.932	1.907	1.829
30	2.338	2.280	2.233	2.195	2.163	2.136	2.074	2.009	1.940	1.892	1.866	1.787
35	2.266	2.207	2.160	2.122	2.089	2.062	1.999	1.932	1.861	1.811	1.785	1.702
40	2.213	2.154	2.107	2.068	2.035	2.007	1.943	1.875	1.803	1.751	1.724	1.637
50	2.140	2.081	2.033	1.993	1.960	1.931	1.866	1.796	1.721	1.667	1.639	1.545
60	2.093	2.033	1.985	1.944	1.911	1.882	1.815	1.744	1.667	1.611	1.581	1.482
70	2.059	1.999	1.950	1.910	1.876	1.847	1.779	1.707	1.628	1.570	1.539	1.436
80	2.035	1.974	1.925	1.884	1.850	1.820	1.752	1.679	1.599	1.540	1.508	1.400
90	2.015	1.955	1.905	1.864	1.830	1.800	1.731	1.657	1.576	1.516	1.483	1.371
120	1.977	1.916	1.866	1.825	1.790	1.760	1.690	1.614	1.530	1.467	1.433	1.310
240	1.921	1.859	1.808	1.766	1.731	1.700	1.628	1.549	1.460	1.392	1.354	1.206
∞	1.866	1.803	1.751	1.708	1.672	1.640	1.566	1.484	1.388	1.313	1.268	1.000

TABLE 1.6

$\alpha = 0.01$ CRITICAL VALUES OF THE F-DISTRIBUTION

	f_1											
$f_2 \downarrow$	1	2	3	4	5	6	7	8	9	10	11	12
6	13.75	10.92	9.780	9.148	8.746	8.466	8.260	8.102	7.976	7.874	7.790	7.718
7	12.25	9.547	8.451	7.847	7.460	7.191	6.993	6.840	6.719	6.620	6.538	6.469
8	11.26	8.649	7.591	7.006	6.632	6.371	6.178	6.029	5.911	5.814	5.734	5.667
9	10.56	8.022	6.992	6.422	6.057	5.802	5.613	5.467	5.351	5.257	5.178	5.111
10	10.04	7.559	6.552	5.994	5.636	5.386	5.200	5.057	4.942	4.849	4.772	4.706
12	9.330	6.927	5.953	5.412	5.064	4.821	4.640	4.499	4.388	4.296	4.220	4.155
14	8.862	6.515	5.564	5.035	4.695	4.456	4.278	4.140	4.030	3.939	3.864	3.800
16	8.531	6.226	5.292	4.773	4.437	4.202	4.026	3.890	3.780	3.691	3.616	3.553
18	8.285	6.013	5.092	4.579	4.248	4.015	3.841	3.705	3.597	3.508	3.434	3.371
20	8.096	5.849	4.938	4.431	4.103	3.871	3.699	3.564	3.457	3.368	3.294	3.231
22	7.945	5.719	4.817	4.313	3.988	3.758	3.587	3.453	3.346	3.258	3.184	3.121
24	7.823	5.614	4.718	4.218	3.895	3.667	3.496	3.363	3.256	3.168	3.094	3.032
26	7.721	5.526	4.637	4.140	3.818	3.591	3.421	3.288	3.182	3.094	3.021	2.958
28	7.636	5.453	4.568	4.074	3.754	3.528	3.358	3.226	3.120	3.032	2.959	2.896
30	7.562	5.390	4.510	4.018	3.699	3.473	3.304	3.173	3.067	2.979	2.906	2.843
35	7.419	5.268	4.396	3.908	3.592	3.368	3.200	3.069	2.963	2.876	2.803	2.740
40	7.314	5.179	4.313	3.828	3.514	3.291	3.124	2.993	2.888	2.801	2.727	2.665
50	7.171	5.057	4.199	3.720	3.408	3.186	3.020	2.890	2.785	2.698	2.625	2.562
60	7.077	4.977	4.126	3.649	3.339	3.119	2.953	2.823	2.718	2.632	2.559	2.496
70	7.011	4.922	4.074	3.600	3.291	3.071	2.906	2.777	2.672	2.585	2.512	2.450
80	6.963	4.881	4.036	3.563	3.255	3.036	2.871	2.742	2.637	2.551	2.478	2.415
90	6.925	4.849	4.007	3.535	3.228	3.009	2.845	2.715	2.611	2.524	2.451	2.389
120	6.851	4.787	3.949	3.480	3.174	2.956	2.792	2.663	2.559	2.472	2.399	2.336
240	6.742	4.695	3.864	3.398	3.094	2.878	2.714	2.586	2.482	2.395	2.322	2.260
∞	6.635	4.605	3.782	3.319	3.017	2.802	2.639	2.511	2.407	2.321	2.248	2.185

	f_1											
$f_2 \downarrow$	14	16	18	20	22	24	30	40	60	90	120	∞
6	7.605	7.519	7.451	7.396	7.351	7.313	7.229	7.143	7.057	6.998	6.969	6.880
7	6.359	6.275	6.209	6.155	6.111	6.074	5.992	5.908	5.824	5.766	5.737	5.650
8	5.559	5.477	5.412	5.359	5.316	5.279	5.198	5.116	5.032	4.975	4.946	4.859
9	5.005	4.924	4.860	4.808	4.765	4.729	4.649	4.567	4.483	4.426	4.398	4.311
10	4.601	4.520	4.457	4.405	4.363	4.327	4.247	4.165	4.082	4.025	3.996	3.909
12	4.052	3.972	3.909	3.858	3.816	3.780	3.701	3.619	3.535	3.478	3.449	3.361
14	3.698	3.619	3.556	3.505	3.463	3.427	3.348	3.266	3.181	3.124	3.094	3.004
16	3.451	3.372	3.310	3.259	3.216	3.181	3.101	3.018	2.933	2.875	2.845	2.753
18	3.269	3.190	3.128	3.077	3.035	2.999	2.919	2.835	2.749	2.690	2.660	2.566
20	3.130	3.051	2.989	2.938	2.895	2.859	2.778	2.695	2.608	2.548	2.517	2.421
22	3.019	2.941	2.879	2.827	2.785	2.749	2.667	2.583	2.495	2.434	2.403	2.305
24	2.930	2.852	2.789	2.738	2.695	2.659	2.577	2.492	2.403	2.342	2.310	2.211
26	2.857	2.778	2.715	2.664	2.621	2.585	2.503	2.417	2.327	2.265	2.233	2.131
28	2.795	2.716	2.653	2.602	2.559	2.522	2.440	2.354	2.263	2.200	2.167	2.064
30	2.742	2.663	2.600	2.549	2.506	2.469	2.386	2.299	2.208	2.144	2.111	2.006
35	2.639	2.560	2.497	2.445	2.401	2.364	2.281	2.193	2.099	2.034	2.000	1.891
40	2.563	2.484	2.421	2.369	2.325	2.288	2.203	2.114	2.019	1.952	1.917	1.805
50	2.461	2.382	2.318	2.265	2.221	2.183	2.098	2.007	1.909	1.839	1.803	1.683
60	2.394	2.315	2.251	2.198	2.153	2.115	2.028	1.936	1.836	1.764	1.726	1.601
70	2.348	2.268	2.204	2.150	2.106	2.067	1.980	1.886	1.785	1.711	1.672	1.540
80	2.313	2.233	2.169	2.115	2.070	2.032	1.944	1.849	1.746	1.671	1.630	1.494
90	2.286	2.206	2.142	2.088	2.043	2.004	1.916	1.820	1.716	1.639	1.598	1.457
120	2.234	2.154	2.089	2.035	1.989	1.950	1.860	1.763	1.656	1.576	1.533	1.381
240	2.157	2.076	2.010	1.956	1.909	1.870	1.778	1.677	1.565	1.480	1.432	1.250
∞	2.082	2.000	1.934	1.878	1.831	1.791	1.696	1.592	1.473	1.379	1.325	1.000

TABLE 1.7

$\alpha = 0.005$ CRITICAL VALUES OF THE F-DISTRIBUTION

	f_1											
$f_2 \downarrow$	1	2	3	4	5	6	7	8	9	10	11	12
6	18.63	14.54	12.92	12.03	11.46	11.07	10.79	10.57	10.39	10.25	10.13	10.03
7	16.24	12.40	10.88	10.05	9.522	9.155	8.885	8.678	8.514	8.380	8.270	8.176
8	14.69	11.04	9.596	8.805	8.302	7.952	7.694	7.496	7.339	7.211	7.104	7.015
9	13.61	10.11	8.717	7.956	7.471	7.134	6.885	6.693	6.541	6.417	6.314	6.227
10	12.83	9.427	8.081	7.343	6.872	6.545	6.302	6.116	5.968	5.847	5.746	5.661
12	11.75	8.510	7.226	6.521	6.071	5.757	5.525	5.345	5.202	5.085	4.988	4.906
14	11.06	7.922	6.680	5.998	5.562	5.257	5.031	4.857	4.717	4.603	4.508	4.428
16	10.58	7.514	6.303	5.638	5.212	4.913	4.692	4.521	4.384	4.272	4.179	4.099
18	10.22	7.215	6.028	5.375	4.956	4.663	4.445	4.276	4.141	4.030	3.938	3.860
20	9.944	6.986	5.818	5.174	4.762	4.472	4.257	4.090	3.956	3.847	3.756	3.678
22	9.727	6.806	5.652	5.017	4.609	4.322	4.109	3.944	3.812	3.703	3.612	3.535
24	9.551	6.661	5.519	4.890	4.486	4.202	3.991	3.826	3.695	3.587	3.497	3.420
26	9.406	6.541	5.409	4.785	4.384	4.103	3.893	3.730	3.599	3.492	3.402	3.325
28	9.284	6.440	5.317	4.698	4.300	4.020	3.811	3.649	3.519	3.412	3.322	3.246
30	9.180	6.355	5.239	4.623	4.228	3.949	3.742	3.580	3.450	3.344	3.255	3.179
35	8.976	6.188	5.086	4.479	4.088	3.812	3.607	3.447	3.318	3.212	3.124	3.048
40	8.828	6.066	4.976	4.374	3.986	3.713	3.509	3.350	3.222	3.117	3.028	2.953
50	8.626	5.902	4.826	4.232	3.849	3.579	3.376	3.219	3.092	2.988	2.900	2.825
60	8.495	5.795	4.729	4.140	3.760	3.492	3.291	3.134	3.008	2.904	2.817	2.742
70	8.403	5.720	4.661	4.076	3.698	3.431	3.232	3.076	2.950	2.846	2.759	2.684
80	8.335	5.665	4.611	4.029	3.652	3.387	3.188	3.032	2.907	2.803	2.716	2.641
90	8.282	5.623	4.573	3.992	3.617	3.352	3.154	2.999	2.873	2.770	2.683	2.608
120	8.179	5.539	4.497	3.921	3.548	3.285	3.087	2.933	2.808	2.705	2.618	2.544
240	8.027	5.417	4.387	3.816	3.447	3.187	2.991	2.837	2.713	2.610	2.524	2.450
∞	7.879	5.298	4.279	3.715	3.350	3.091	2.897	2.744	2.621	2.519	2.432	2.358

	f_1											
$f_2 \downarrow$	14	16	18	20	22	24	30	40	60	90	120	∞
6	9.877	9.758	9.664	9.589	9.526	9.474	9.358	9.241	9.122	9.042	9.001	8.879
7	8.028	7.915	7.826	7.754	7.695	7.645	7.534	7.422	7.309	7.232	7.193	7.076
8	6.872	6.763	6.678	6.608	6.551	6.503	6.396	6.288	6.177	6.103	6.065	5.951
9	6.089	5.983	5.899	5.832	5.776	5.729	5.625	5.519	5.410	5.337	5.300	5.188
10	5.526	5.422	5.340	5.274	5.219	5.173	5.071	4.966	4.859	4.787	4.750	4.639
12	4.775	4.674	4.595	4.530	4.476	4.431	4.331	4.228	4.123	4.051	4.015	3.904
14	4.299	4.200	4.122	4.059	4.006	3.961	3.862	3.760	3.655	3.584	3.547	3.436
16	3.972	3.875	3.797	3.734	3.682	3.638	3.539	3.437	3.332	3.261	3.224	3.112
18	3.734	3.637	3.560	3.498	3.446	3.402	3.303	3.201	3.096	3.024	2.987	2.873
20	3.553	3.457	3.380	3.318	3.266	3.222	3.123	3.022	2.916	2.843	2.806	2.690
22	3.411	3.315	3.239	3.176	3.125	3.081	2.982	2.880	2.774	2.700	2.663	2.545
24	3.296	3.201	3.125	3.062	3.011	2.967	2.868	2.765	2.658	2.584	2.546	2.428
26	3.202	3.107	3.031	2.968	2.917	2.873	2.774	2.671	2.563	2.489	2.450	2.330
28	3.123	3.028	2.952	2.890	2.838	2.794	2.695	2.592	2.483	2.408	2.369	2.247
30	3.056	2.961	2.885	2.823	2.771	2.727	2.628	2.524	2.415	2.339	2.300	2.176
35	2.926	2.831	2.755	2.693	2.641	2.597	2.497	2.392	2.282	2.204	2.164	2.036
40	2.831	2.737	2.661	2.598	2.546	2.502	2.401	2.296	2.184	2.105	2.064	1.932
50	2.703	2.609	2.533	2.470	2.418	2.373	2.272	2.164	2.050	1.968	1.925	1.786
60	2.620	2.526	2.450	2.387	2.335	2.290	2.187	2.079	1.962	1.878	1.834	1.689
70	2.563	2.468	2.392	2.329	2.276	2.231	2.128	2.019	1.900	1.815	1.769	1.618
80	2.520	2.425	2.349	2.286	2.233	2.188	2.084	1.974	1.854	1.767	1.720	1.563
90	2.487	2.393	2.316	2.253	2.200	2.155	2.051	1.939	1.818	1.730	1.682	1.520
120	2.423	2.328	2.251	2.188	2.135	2.089	1.984	1.871	1.747	1.655	1.606	1.431
240	2.329	2.233	2.157	2.093	2.039	1.993	1.886	1.770	1.640	1.542	1.488	1.281
∞	2.237	2.142	2.064	2.000	1.945	1.898	1.789	1.669	1.533	1.426	1.364	1.000

TABLE 1.8

$\alpha = 0.001$ CRITICAL VALUES OF THE F-DISTRIBUTION

	f_1											
$f_2 \downarrow$	1	2	3	4	5	6	7	8	9	10	11	12
6	35.51	27.00	23.70	21.92	20.80	20.03	19.46	19.03	18.69	18.41	18.18	17.99
7	29.25	21.69	18.77	17.20	16.21	15.52	15.02	14.63	14.33	14.08	13.88	13.71
8	25.41	18.49	15.83	14.39	13.48	12.86	12.40	12.05	11.77	11.54	11.35	11.19
9	22.86	16.39	13.90	12.56	11.71	11.13	10.70	10.37	10.11	9.894	9.718	9.570
10	21.04	14.91	12.55	11:28	10.48	9.926	9.517	9.204	8.956	8.754	8.586	8.445
12	18.64	12.97	10.80	9.633	8.892	8.379	8.001	7.710	7.480	7.292	7.136	7.005
14	17.14	11.78	9.729	8.622	7.922	7.436	7.077	6.802	6.583	6.404	6.256	6.130
16	16.12	10.97	9.006	7.944	7.272	6.805	6.460	6.195	5.984	5.812	5.668	5.547
18	15.38	10.39	8.487	7.459	6.808	6.355	6.021	5.763	5.558	5.390	5.250	5.132
20	14.82	9.953	8.098	7.096	6.461	6.019	5.692	5.440	5.239	5.075	4.939	4.823
22	14.38	9.612	7.796	6.814	6.191	5.758	5.438	5.190	4.993	4.832	4.697	4.583
24	14.03	9.339	7.554	6.589	5.977	5.550	5.235	4.991	4.797	4.638	4.505	4.393
26	13.74	9.116	7.357	6.406	5.802	5.381	5.070	4.829	4.637	4.480	4.349	4.238
28	13.50	8.931	7.193	6.253	5.656	5.241	4.933	4.695	4.505	4.349	4.219	4.109
30	13.29	8.773	7.054	6.125	5.534	5.122	4.817	4.581	4.393	4.239	4.110	4.001
35	12.90	8.470	6.787	5.876	5.298	4.894	4.595	4.363	4.178	4.027	3.900	3.792
40	12.61	8.251	6.595	5.698	5.128	4.731	4.436	4.207	4.024	3.874	3.749	3.642
50	12.22	7.956	6.336	5.459	4.901	4.512	4.222	3.998	3.818	3.671	3.548	3.443
60	11.97	7.768	6.171	5.307	4.757	4.372	4.086	3.865	3.687	3.541	3.419	3.315
70	11.80	7.637	6.057	5.201	4.656	4.275	3.992	3.773	3.596	3.452	3.330	3.227
80	11.67	7.540	5.972	5.123	4.582	4.204	3.923	3.705	3.530	3.386	3.265	3.162
90	11.57	7.466	5.908	5.064	4.526	4.150	3.870	3.653	3.479	3.336	3.215	3.113
120	11.38	7.321	5.781	4.947	4.416	4.044	3.767	3.552	3.379	3.237	3.118	3.016
240	11.10	7.110	5.598	4.778	4.256	3.890	3.618	3.406	3.235	3.095	2.977	2.876
∞	10.83	6.908	5.422	4.617	4.103	3.743	3.475	3.266	3.097	2.959	2.842	2.742

$f_2 \downarrow$	f_1 14	16	18	20	22	24	30	40	60	90	120	∞
6	17.68	17.45	17.27	17.12	17.00	16.90	16.67	16.44	16.21	16.06	15.98	15.75
7	13.43	13.23	13.06	12.93	12.82	12.73	12.53	12.33	12.12	11.98	11.91	11.70
8	10.94	10.75	10.60	10.48	10.38	10.30	10.11	9.919	9.727	9.597	9.532	9.334
9	9.334	9.154	9.012	8.898	8.803	8.724	8.548	8.369	8.187	8.063	8.001	7.813
10	8.220	8.048	7.913	7.804	7.713	7.638	7.469	7.297	7.122	7.004	6.944	6.762
12	6.794	6.634	6.507	6.405	6.320	6.249	6.090	5.928	5.762	5.650	5.593	5.420
14	5.930	5.776	5.655	5.557	5.475	5.407	5.254	5.098	4.938	4.829	4.773	4.604
16	5.353	5.205	5.087	4.992	4.913	4.846	4.697	4.545	4.388	4.281	4.226	4.059
18	4.943	4.798	4.683	4.590	4.512	4.447	4.301	4.151	3.996	3.890	3.836	3.670
20	4.637	4.495	4.382	4.290	4.214	4.149	4.005	3.856	3.703	3.598	3.544	3.378
22	4.401	4.260	4.149	4.058	3.983	3.919	3.776	3.629	3.476	3.371	3.317	3.151
24	4.212	4.074	3.963	3.873	3.799	3.735	3.593	3.447	3.295	3.190	3.136	2.969
26	4.059	3.921	3.812	3.723	3.649	3.586	3.445	3.299	3.147	3.041	2.988	2.819
28	3.932	3.795	3.687	3.598	3.524	3.462	3.321	3.176	3.024	2.918	2.864	2.695
30	3.825	3.689	3.581	3.493	3.419	3.357	3.217	3.072	2.920	2.814	2.760	2.589
35	3.618	3.484	3.378	3.290	3.217	3.156	3.016	2.871	2.719	2.612	2.557	2.383
40	3.471	3.338	3.232	3.145	3.073	3.011	2.872	2.727	2.574	2.466	2.410	2.233
50	3.273	3.142	3.037	2.951	2.879	2.817	2.679	2.533	2.378	2.269	2.211	2.026
60	3.147	3.017	2.912	2.827	2.755	2.694	2.555	2.409	2.252	2.141	2.082	1.890
70	3.060	2.930	2.826	2.741	2.669	2.608	2.469	2.322	2.164	2.051	1.991	1.793
80	2.996	2.867	2.763	2.677	2.606	2.545	2.406	2.258	2.099	1.984	1.924	1.720
90	2.947	2.818	2.714	2.629	2.558	2.497	2.357	2.209	2.049	1.933	1.871	1.662
120	2.851	2.723	2.620	2.534	2.463	2.402	2.262	2.113	1.950	1.831	1.767	1.543
240	2.713	2.585	2.482	2.397	2.326	2.265	2.124	1.972	1.804	1.679	1.609	1.349
∞	2.580	2.453	2.351	2.266	2.194	2.132	1.990	1.835	1.660	1.525	1.447	1.000

TABLE 2.1
CRITICAL VALUES OF THE χ^2-DISTRIBUTION

	α							
$\nu \downarrow$	0.500	0.250	0.100	0.050	0.025	0.010	0.005	0.001
1	0.45494	1.32330	2.70554	3.84146	5.02389	6.63490	7.87944	10.8276
2	1.38629	2.77259	4.60517	5.99146	7.37776	9.21034	10.5966	13.8155
3	2.36597	4.10834	6.25139	7.81473	9.34840	11.3449	12.8382	16.2662
4	3.35669	5.38527	7.77944	9.48773	11.1433	13.2767	14.8603	18.4668
5	4.35146	6.62568	9.23636	11.0705	12.8325	15.0863	16.7496	20.5150
6	5.34812	7.84080	10.6446	12.5916	14.4494	16.8119	18.5476	22.4577
7	6.34581	9.03715	12.0170	14.0671	16.0128	18.4753	20.2777	24.3219
8	7.34412	10.2189	13.3616	15.5073	17.5345	20.0902	21.9550	26.1245
9	8.34283	11.3888	14.6837	16.9190	19.0228	21.6660	23.5894	27.8772
10	9.34182	12.5489	15.9872	18.3070	20.4832	23.2093	25.1882	29.5883
11	10.3410	13.7007	17.2750	19.6751	21.9200	24.7250	26.7568	31.2641
12	11.3403	14.8454	18.5493	21.0261	23.3367	26.2170	28.2995	32.9095
13	12.3398	15.9839	19.8119	22.3620	24.7356	27.6882	29.8195	34.5282
14	13.3393	17.1169	21.0641	23.6848	26.1189	29.1412	31.3193	36.1233
15	14.3389	18.2451	22.3071	24.9958	27.4884	30.5779	32.8013	37.6973
16	15.3385	19.3689	23.5418	26.2962	28.8454	31.9999	34.2672	39.2524
17	16.3382	20.4887	24.7690	27.5871	30.1910	33.4087	35.7185	40.7902
18	17.3379	21.6049	25.9894	28.8693	31.5264	34.8053	37.1565	42.3124
19	18.3377	22.7178	27.2036	30.1435	32.8523	36.1909	38.5823	43.8202
20	19.3374	23.8277	28.4120	31.4104	34.1696	37.5662	39.9968	45.3147
21	20.3372	24.9348	29.6151	32.6706	35.4789	38.9322	41.4011	46.7970
22	21.3370	26.0393	30.8133	33.9244	36.7807	40.2894	42.7957	48.2679
23	22.3369	27.1413	32.0069	35.1725	38.0756	41.6384	44.1813	49.7282
24	23.3367	28.2412	33.1962	36.4150	39.3641	42.9798	45.5585	51.1786
25	24.3366	29.3389	34.3816	37.6525	40.6465	44.3141	46.9279	52.6197
26	25.3365	30.4346	35.5632	38.8851	41.9232	45.6417	48.2899	54.0520
27	26.3363	31.5284	36.7412	40.1133	43.1945	46.9629	49.6449	55.4760
28	27.3362	32.6205	37.9159	41.3371	44.4608	48.2782	50.9934	56.8923
29	28.3361	33.7109	39.0875	42.5570	45.7223	49.5879	52.3356	58.3012
30	29.3360	34.7997	40.2560	43.7730	46.9792	50.8922	53.6720	59.7031
31	30.3359	35.8871	41.4217	44.9853	48.2319	52.1914	55.0027	61.0983
32	31.3359	36.9730	42.5847	46.1943	49.4804	53.4858	56.3281	62.4872
33	32.3358	38.0575	43.7452	47.3999	50.7251	54.7755	57.6484	63.8701
34	33.3357	39.1408	44.9032	48.6024	51.9660	56.0609	58.9639	65.2472
35	34.3356	40.2228	46.0588	49.8018	53.2033	57.3421	60.2748	66.6188
36	35.3356	41.3036	47.2122	50.9985	54.4373	58.6192	61.5812	67.9852
37	36.3355	42.3833	48.3634	52.1923	55.6680	59.8925	62.8833	69.3465
38	37.3355	43.4619	49.5126	53.3835	56.8955	61.1621	64.1814	70.7029
39	38.3354	44.5395	50.6598	54.5722	58.1201	62.4281	65.4756	72.0547
40	39.3353	45.6160	51.8051	55.7585	59.3417	63.6907	66.7660	73.4020
41	40.3353	46.6916	52.9485	56.9424	60.5606	64.9501	68.0527	74.7449
42	41.3352	47.7663	54.0902	58.1240	61.7768	66.2062	69.3360	76.0838
43	42.3352	48.8400	55.2302	59.3035	62.9904	67.4593	70.6159	77.4186
44	43.3352	49.9129	56.3685	60.4809	64.2015	68.7095	71.8926	78.7495
45	44.3351	50.9849	57.5053	61.6562	65.4102	69.9568	73.1661	80.0767

TABLE 2.2
CRITICAL VALUES OF THE χ^2-DISTRIBUTION

	α							
$\nu \downarrow$	0.500	0.250	0.100	0.050	0.025	0.010	0.005	0.001
46	45.3351	52.0562	58.6405	62.8296	66.6165	71.2014	74.4365	81.4003
47	46.3350	53.1267	59.7743	64.0011	67.8206	72.4433	75.7041	82.7204
48	47.3350	54.1964	60.9066	65.1708	69.0226	73.6826	76.9688	84.0371
49	48.3350	55.2653	62.0375	66.3386	70.2224	74.9195	78.2307	85.3506
50	49.3349	56.3336	63.1671	67.5048	71.4202	76.1539	79.4900	86.6608
51	50.3349	57.4012	64.2954	68.6693	72.6160	77.3860	80.7467	87.9680
52	51.3349	58.4681	65.4224	69.8322	73.8099	78.6158	82.0008	89.2722
53	52.3348	59.5344	66.5482	70.9935	75.0019	79.8433	83.2526	90.5734
54	53.3348	60.6000	67.6728	72.1532	76.1920	81.0688	84.5019	91.8718
55	54.3348	61.6650	68.7962	73.3115	77.3805	82.2921	85.7490	93.1675
56	55.3348	62.7294	69.9185	74.4683	78.5672	83.5134	86.9938	94.4605
57	56.3347	63.7933	71.0397	75.6237	79.7522	84.7328	88.2364	95.7510
58	57.3347	64.8565	72.1598	76.7778	80.9356	85.9502	89.4769	97.0388
59	58.3347	65.9193	73.2789	77.9305	82.1174	87.1657	90.7153	98.3242
60	59.3347	66.9815	74.3970	79.0819	83.2977	88.3794	91.9517	99.6072
62	61.3346	69.1043	76.6302	81.3810	85.6537	90.8015	94.4187	102.166
64	63.3346	71.2251	78.8596	83.6753	88.0041	93.2169	96.8781	104.716
66	65.3345	73.3441	81.0855	85.9649	90.3489	95.6257	99.3304	107.258
68	67.3345	75.4612	83.3079	88.2502	92.6885	98.0284	101.776	109.791
70	69.3345	77.5767	85.5270	90.5312	95.0232	100.425	104.215	112.317
75	74.3344	82.8581	91.0615	96.2167	100.839	106.393	110.286	118.599
80	79.3343	88.1303	96.5782	101.879	106.629	112.329	116.321	124.839
85	84.3343	93.3939	102.079	107.522	112.393	118.236	122.325	131.041
90	89.3342	98.6499	107.565	113.145	118.136	124.116	128.299	137.208
95	94.3342	103.899	113.038	118.752	123.858	129.973	134.247	143.344
100	99.3341	109.141	118.498	124.342	129.561	135.807	140.169	149.449
105	104.334	114.378	123.947	129.918	135.247	141.620	146.070	155.528
110	109.334	119.608	129.385	135.480	140.917	147.414	151.948	161.581
115	114.334	124.834	134.813	141.030	146.571	153.191	157.808	167.610
120	119.334	130.055	140.233	146.567	152.211	158.950	163.648	173.617
125	124.334	135.271	145.643	152.094	157.839	164.694	169.471	179.604
130	129.334	140.482	151.045	157.610	163.453	170.423	175.278	185.571
135	134.334	145.690	156.440	163.116	169.056	176.138	181.070	191.520
140	139.334	150.894	161.827	168.613	174.648	181.840	186.847	197.451
145	144.334	156.094	167.207	174.101	180.229	187.530	192.610	203.366
150	149.334	161.291	172.581	179.581	185.800	193.208	198.360	209.265
155	154.334	166.485	177.949	185.052	191.362	198.874	204.098	215.149
160	159.334	171.675	183.311	190.516	196.915	204.530	209.824	221.019
165	164.334	176.863	188.667	195.973	202.459	210.176	215.539	226.876
170	169.334	182.047	194.017	201.423	207.995	215.812	221.242	232.719
175	174.334	187.229	199.363	206.867	213.524	221.438	226.936	238.551
180	179.334	192.409	204.704	212.304	219.044	227.056	232.620	244.370
185	184.334	197.586	210.040	217.735	224.558	232.665	238.294	250.179
190	189.334	202.760	215.371	223.160	230.064	238.266	243.959	255.976
200	199.334	213.102	226.021	233.994	241.058	249.445	255.264	267.541

TABLE 3

CRITICAL VALUES OF THE t-DISTRIBUTION

	α								
$\nu \downarrow$	0.2500	0.1000	0.0500	0.0250	0.0100	0.0050	0.0025	0.0010	0.0005
1	1.0000	3.0777	6.3138	12.706	31.821	63.657	127.32	318.31	636.62
2	0.8165	1.8856	2.9200	4.3027	6.9646	9.9248	14.089	22.327	31.599
3	0.7649	1.6377	2.3534	3.1824	4.5407	5.8409	7.4533	10.215	12.924
4	0.7407	1.5332	2.1318	2.7764	3.7469	4.6041	5.5976	7.1732	8.6103
5	0.7267	1.4759	2.0150	2.5706	3.3649	4.0321	4.7733	5.8934	6.8688
6	0.7176	1.4398	1.9432	2.4469	3.1427	3.7074	4.3168	5.2076	5.9588
7	0.7111	1.4149	1.8946	2.3646	2.9980	3.4995	4.0293	4.7853	5.4079
8	0.7064	1.3968	1.8595	2.3060	2.8965	3.3554	3.8325	4.5008	5.0413
9	0.7027	1.3830	1.8331	2.2622	2.8214	3.2498	3.6897	4.2968	4.7809
10	0.6998	1.3722	1.8125	2.2281	2.7638	3.1693	3.5814	4.1437	4.5869
11	0.6974	1.3634	1.7959	2.2010	2.7181	3.1058	3.4966	4.0247	4.4370
12	0.6955	1.3562	1.7823	2.1788	2.6810	3.0545	3.4284	3.9296	4.3178
13	0.6938	1.3502	1.7709	2.1604	2.6503	3.0123	3.3725	3.8520	4.2208
14	0.6924	1.3450	1.7613	2.1448	2.6245	2.9768	3.3257	3.7874	4.1405
15	0.6912	1.3406	1.7531	2.1314	2.6025	2.9467	3.2860	3.7328	4.0728
16	0.6901	1.3368	1.7459	2.1199	2.5835	2.9208	3.2520	3.6862	4.0150
17	0.6892	1.3334	1.7396	2.1098	2.5669	2.8982	3.2224	3.6458	3.9651
18	0.6884	1.3304	1.7341	2.1009	2.5524	2.8784	3.1966	3.6105	3.9216
19	0.6876	1.3277	1.7291	2.0930	2.5395	2.8609	3.1737	3.5794	3.8834
20	0.6870	1.3253	1.7247	2.0860	2.5280	2.8453	3.1534	3.5518	3.8495
21	0.6864	1.3232	1.7207	2.0796	2.5176	2.8314	3.1352	3.5272	3.8193
22	0.6858	1.3212	1.7171	2.0739	2.5083	2.8188	3.1188	3.5050	3.7921
23	0.6853	1.3195	1.7139	2.0687	2.4999	2.8073	3.1040	3.4850	3.7676
24	0.6848	1.3178	1.7109	2.0639	2.4922	2.7969	3.0905	3.4668	3.7454
25	0.6844	1.3163	1.7081	2.0595	2.4851	2.7874	3.0782	3.4502	3.7251
26	0.6840	1.3150	1.7056	2.0555	2.4786	2.7787	3.0669	3.4350	3.7066
27	0.6837	1.3137	1.7033	2.0518	2.4727	2.7707	3.0565	3.4210	3.6896
28	0.6834	1.3125	1.7011	2.0484	2.4671	2.7633	3.0469	3.4082	3.6739
29	0.6830	1.3114	1.6991	2.0452	2.4620	2.7564	3.0380	3.3962	3.6594
30	0.6828	1.3104	1.6973	2.0423	2.4573	2.7500	3.0298	3.3852	3.6460
40	0.6807	1.3031	1.6839	2.0211	2.4233	2.7045	2.9712	3.3069	3.5510
50	0.6794	1.2987	1.6759	2.0086	2.4033	2.6778	2.9370	3.2614	3.4960
60	0.6786	1.2958	1.6706	2.0003	2.3901	2.6603	2.9146	3.2317	3.4602
70	0.6780	1.2938	1.6669	1.9944	2.3808	2.6479	2.8987	3.2108	3.4350
80	0.6776	1.2922	1.6641	1.9901	2.3739	2.6387	2.8870	3.1953	3.4163
90	0.6772	1.2910	1.6620	1.9867	2.3685	2.6316	2.8779	3.1833	3.4019
100	0.6770	1.2901	1.6602	1.9840	2.3642	2.6259	2.8707	3.1737	3.3905
120	0.6765	1.2886	1.6577	1.9799	2.3578	2.6174	2.8599	3.1595	3.3735
140	0.6762	1.2876	1.6558	1.9771	2.3533	2.6114	2.8522	3.1495	3.3614
160	0.6760	1.2869	1.6544	1.9749	2.3499	2.6069	2.8465	3.1419	3.3524
180	0.6759	1.2863	1.6534	1.9732	2.3472	2.6034	2.8421	3.1361	3.3454
200	0.6757	1.2858	1.6525	1.9719	2.3451	2.6006	2.8385	3.1315	3.3398
300	0.6753	1.2844	1.6499	1.9679	2.3388	2.5923	2.8279	3.1176	3.3233
400	0.6751	1.2837	1.6487	1.9659	2.3357	2.5882	2.8227	3.1107	3.3150
500	0.6750	1.2832	1.6479	1.9647	2.3338	2.5857	2.8195	3.1066	3.3101
∞	0.6745	1.2816	1.6449	1.9600	2.3263	2.5758	2.8070	3.0902	3.2905

PART THREE

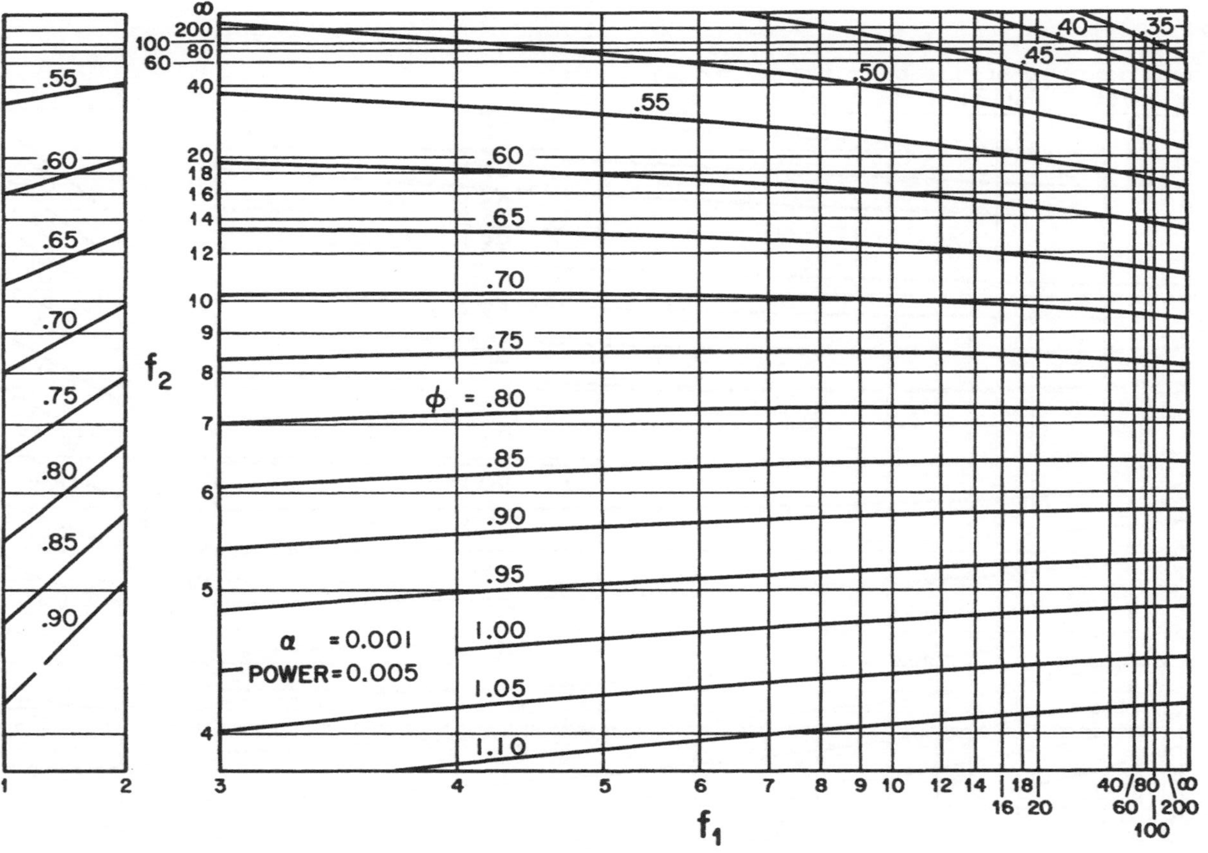

α = 0.001
POWER = 0.005
φ = .80
.35
.40
.45
.50
.55
.60
.65
.70
.75
.85
.90
.95
1.00
1.05
1.10
f_1
f_2

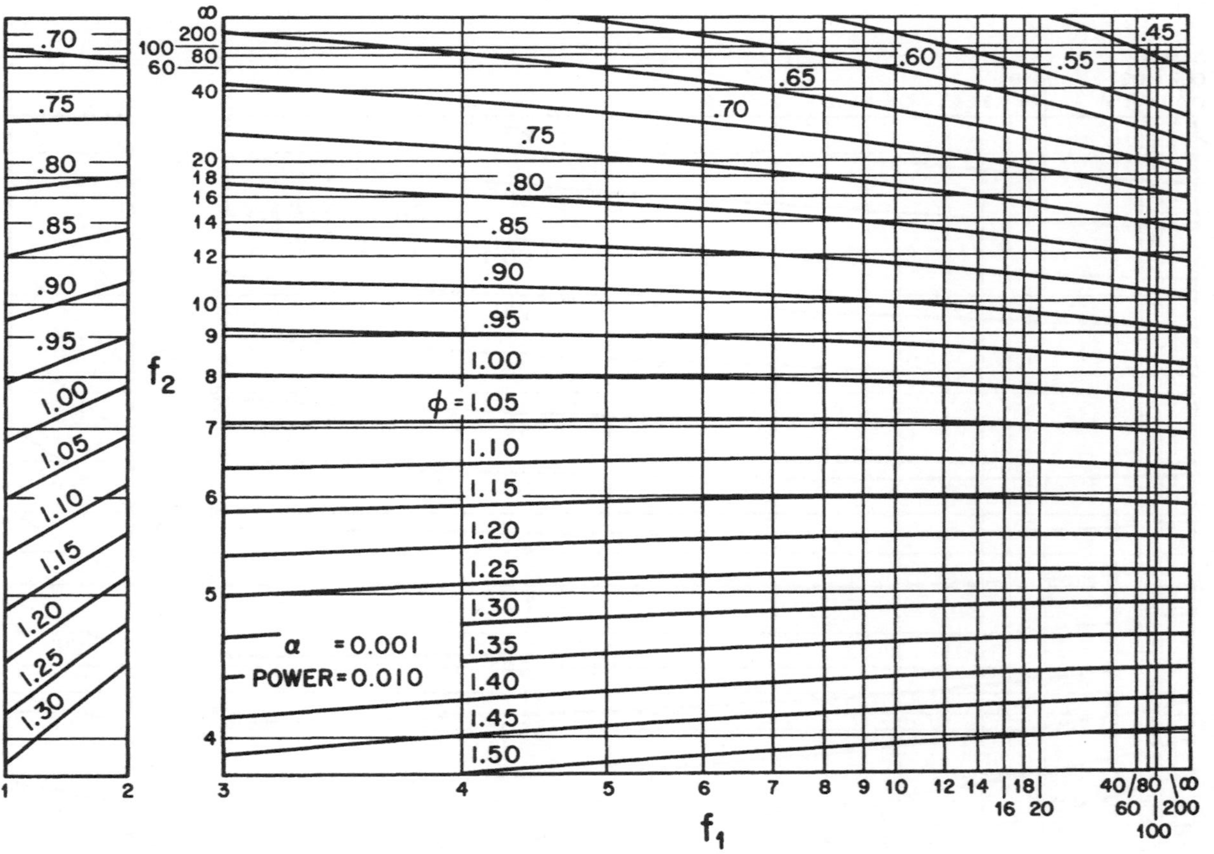

α = 0.001
POWER = 0.010
φ = 1.05
.45
.55
.60
.65
.70
.75
.80
.85
.90
.95
1.00
1.10
1.15
1.20
1.25
1.30
1.35
1.40
1.45
1.50
f1
f2

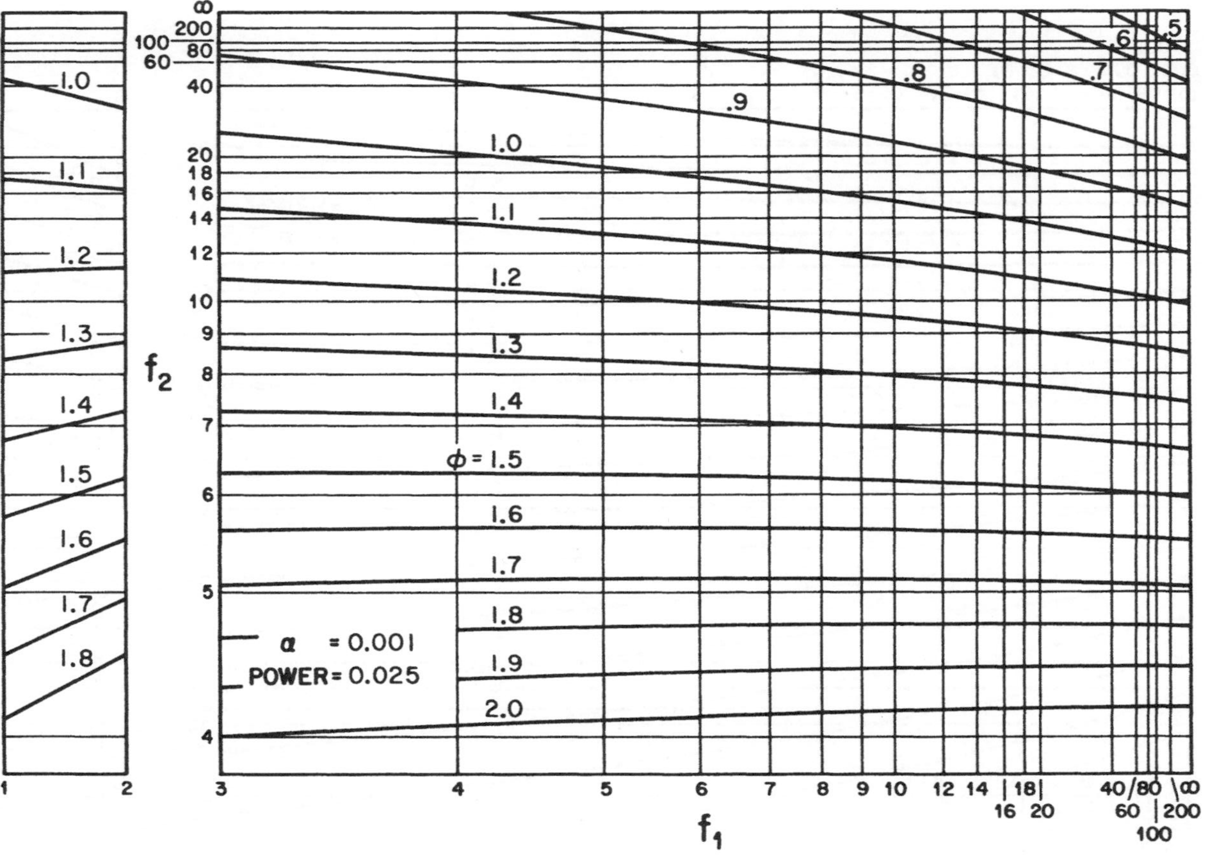

α = 0.001
POWER = 0.025
φ = 1.5
1.0
1.1
1.2
1.3
1.4
1.6
1.7
1.8
1.9
2.0
.9
.8
.7
.6
.5
f_1
f_2

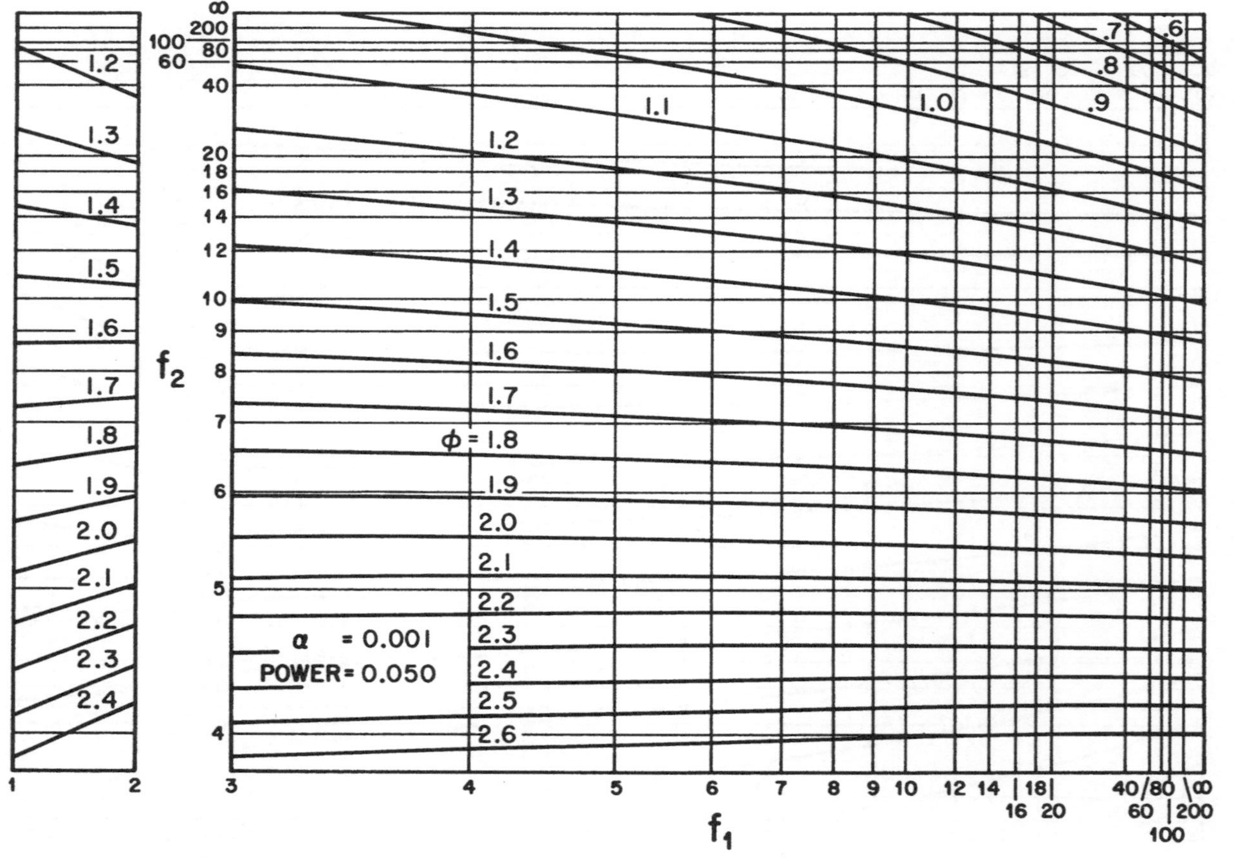
α = 0.001
POWER = 0.050
φ = 1.8
f_1
f_2
1 2 3 4 5 6 7 8 9 10 12 14 16 18 20 40 60 80 100 200 ∞
.6 .7 .8 .9 1.0 1.1 1.2 1.3 1.4 1.5 1.6 1.7 1.8 1.9 2.0 2.1 2.2 2.3 2.4 2.5 2.6

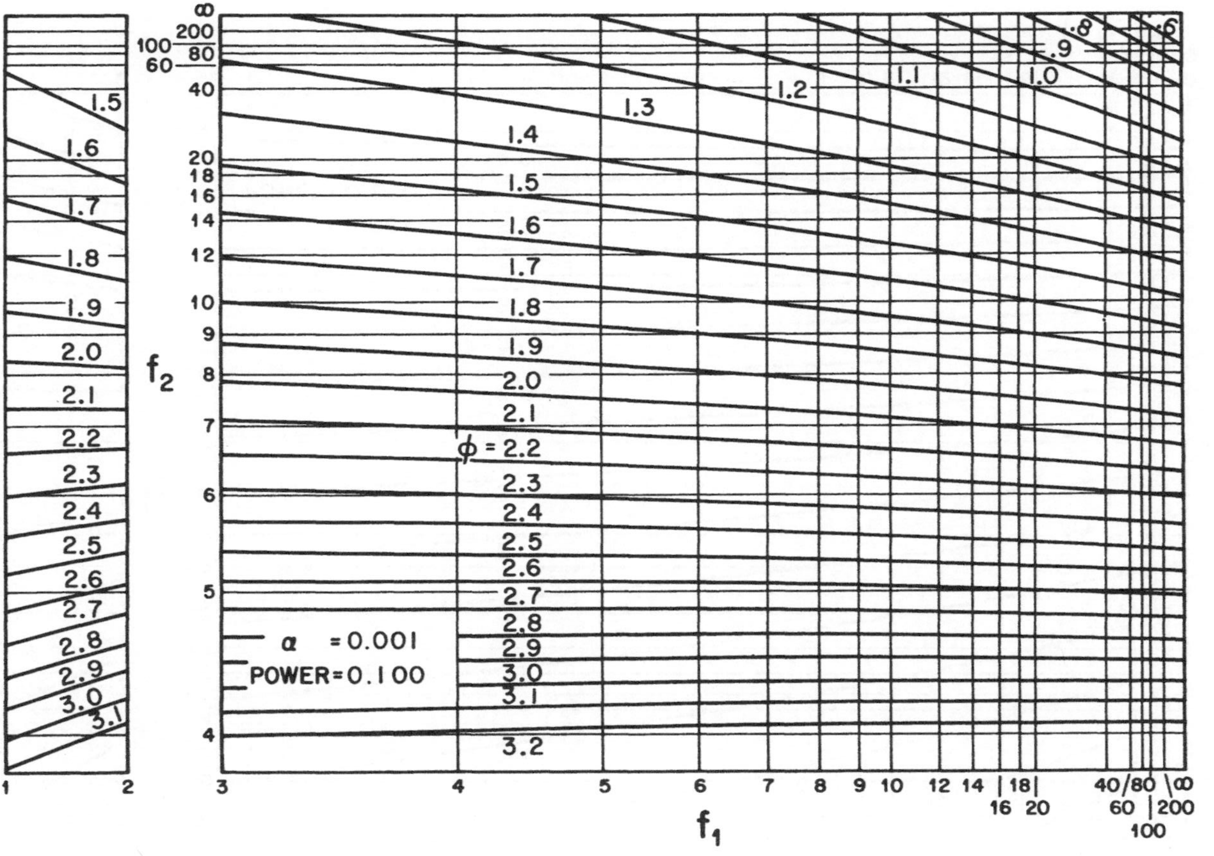

α = 0.001
POWER = 0.100
φ = 2.2
f_1
f_2

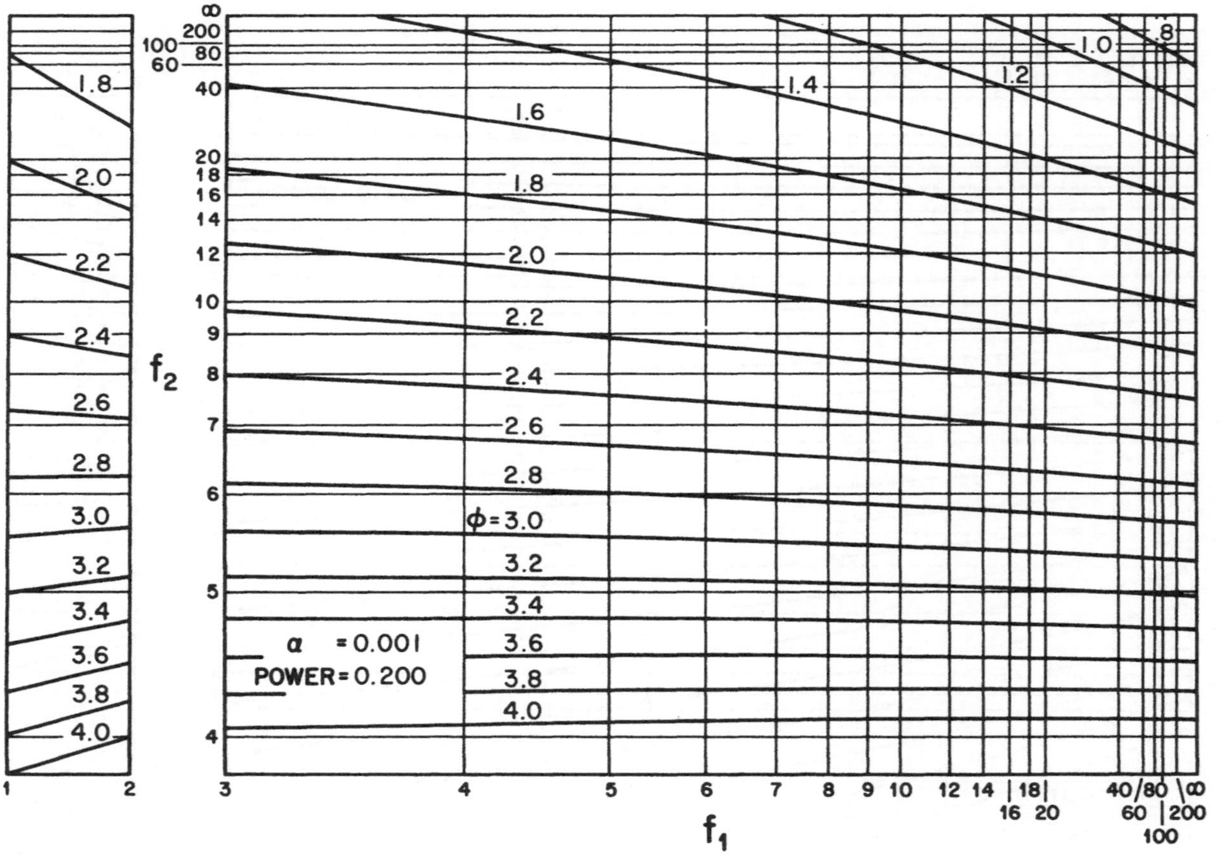

α = 0.001
POWER = 0.200
f_1
f_2
φ = 3.0

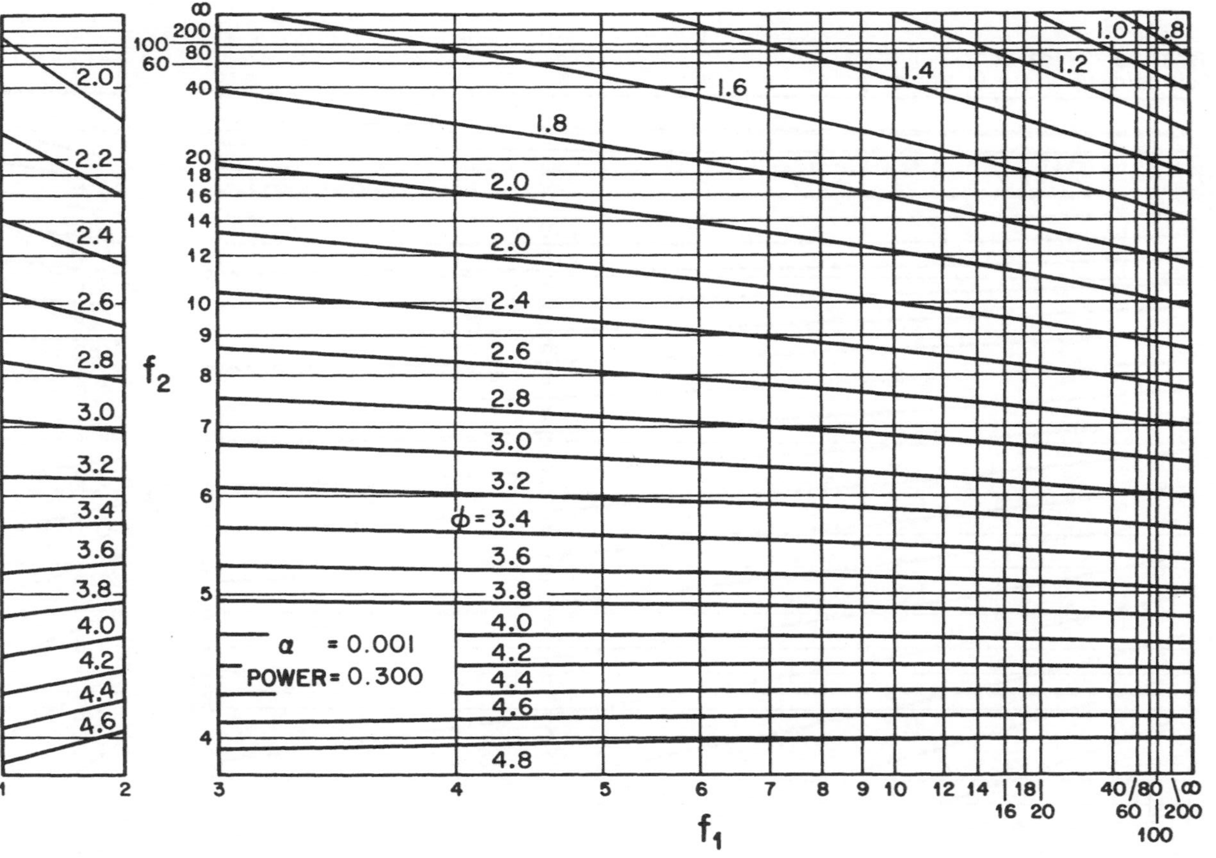
α = 0.001
POWER = 0.300
φ = 3.4
f1
f2

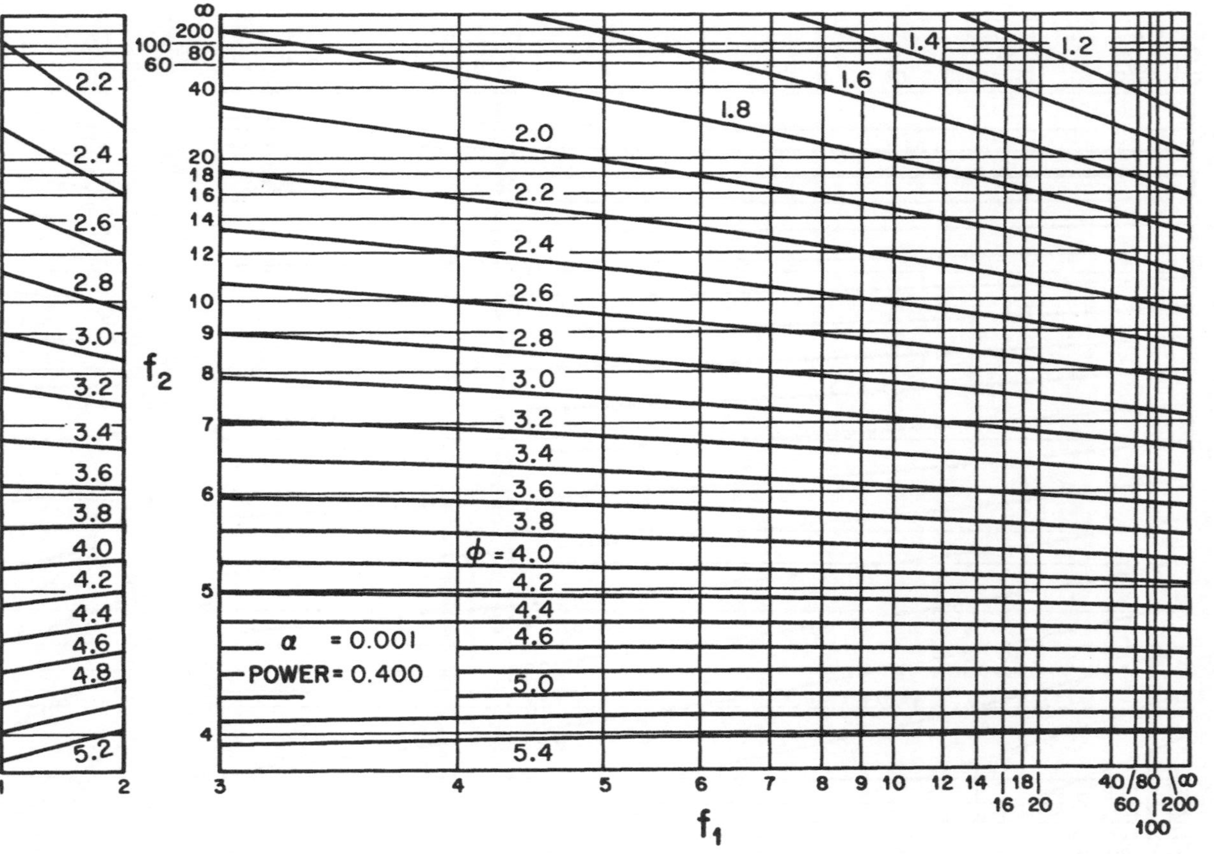
α = 0.001
POWER = 0.400
φ = 4.0
f_1
f_2

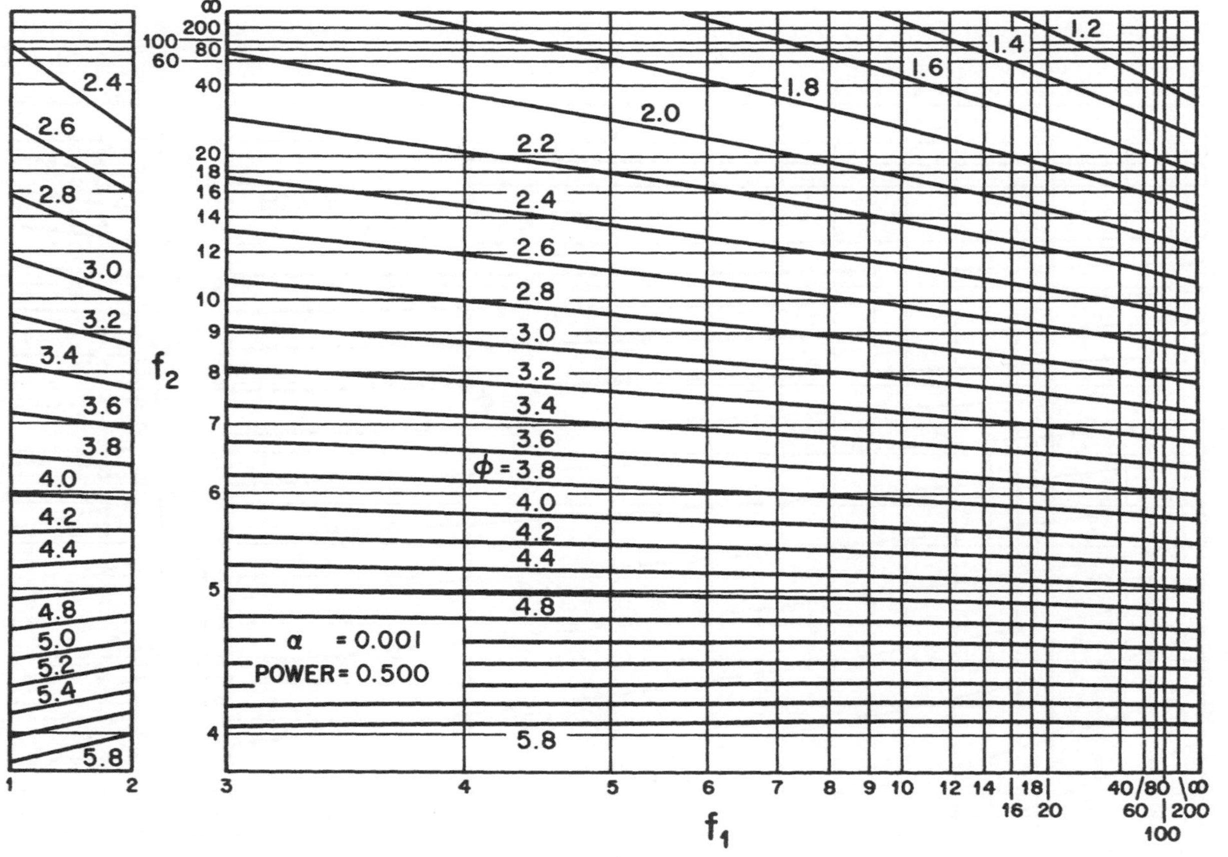
α = 0.001
POWER = 0.500
φ = 3.8
f1
f2

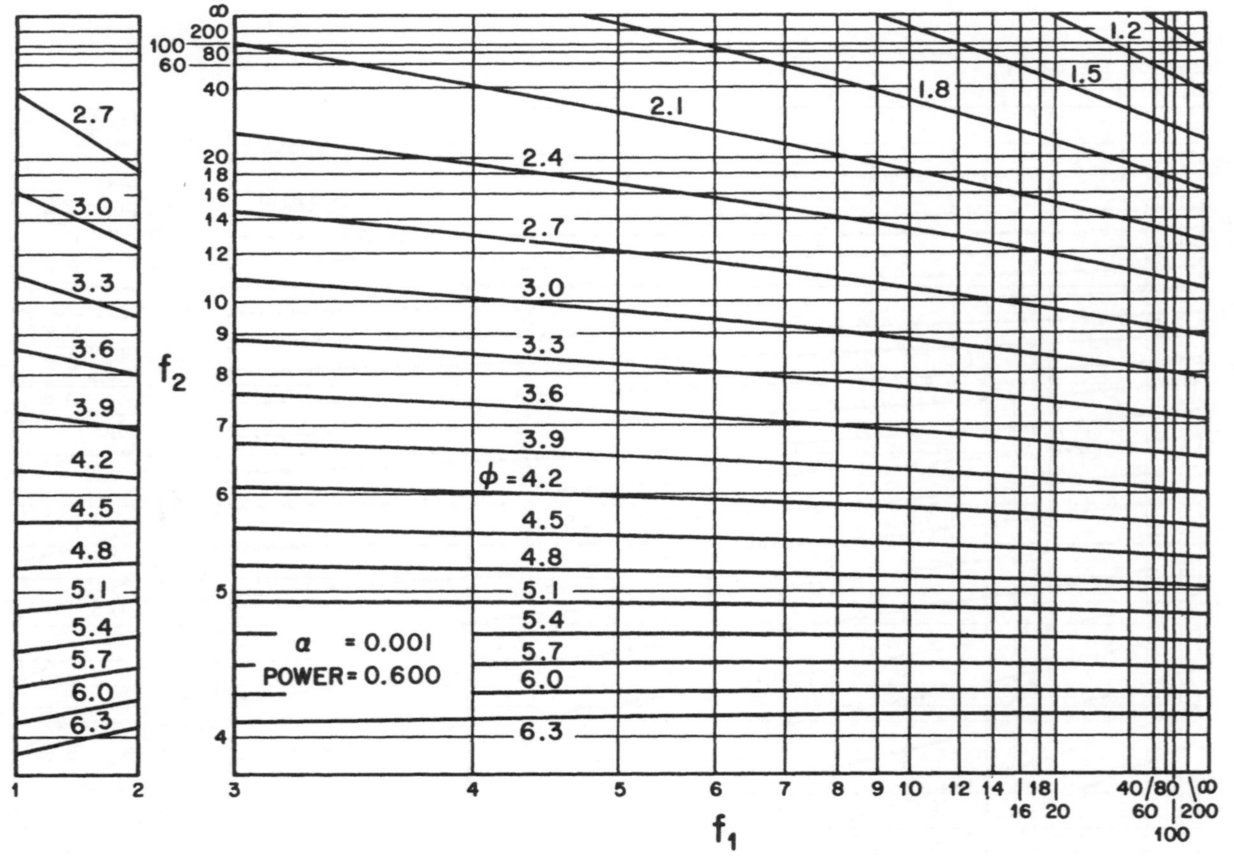

α = 0.001
POWER = 0.600
ϕ = 4.2
f_1
f_2
1.2
1.5
1.8
2.1
2.4
2.7
3.0
3.3
3.6
3.9
4.2
4.5
4.8
5.1
5.4
5.7
6.0
6.3

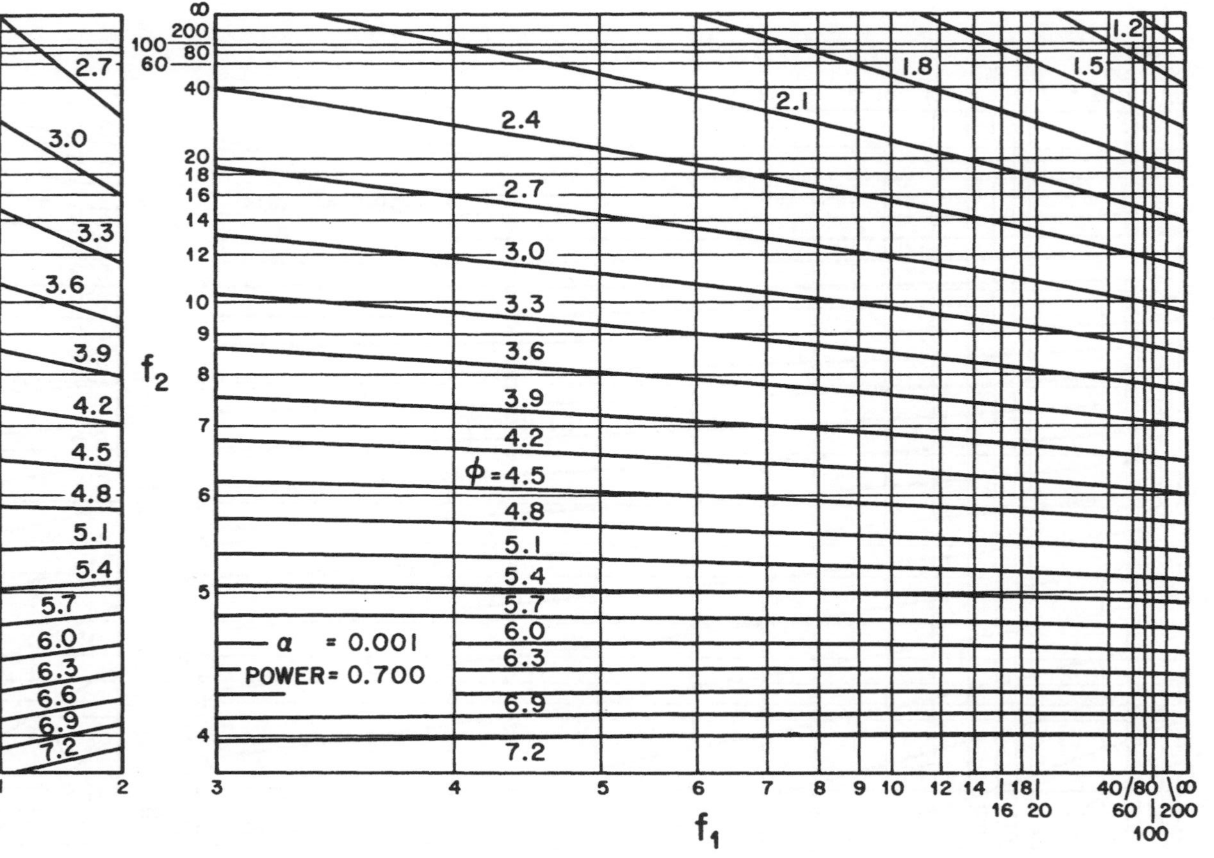

α = 0.001
POWER = 0.700
φ = 4.5
f_1
f_2

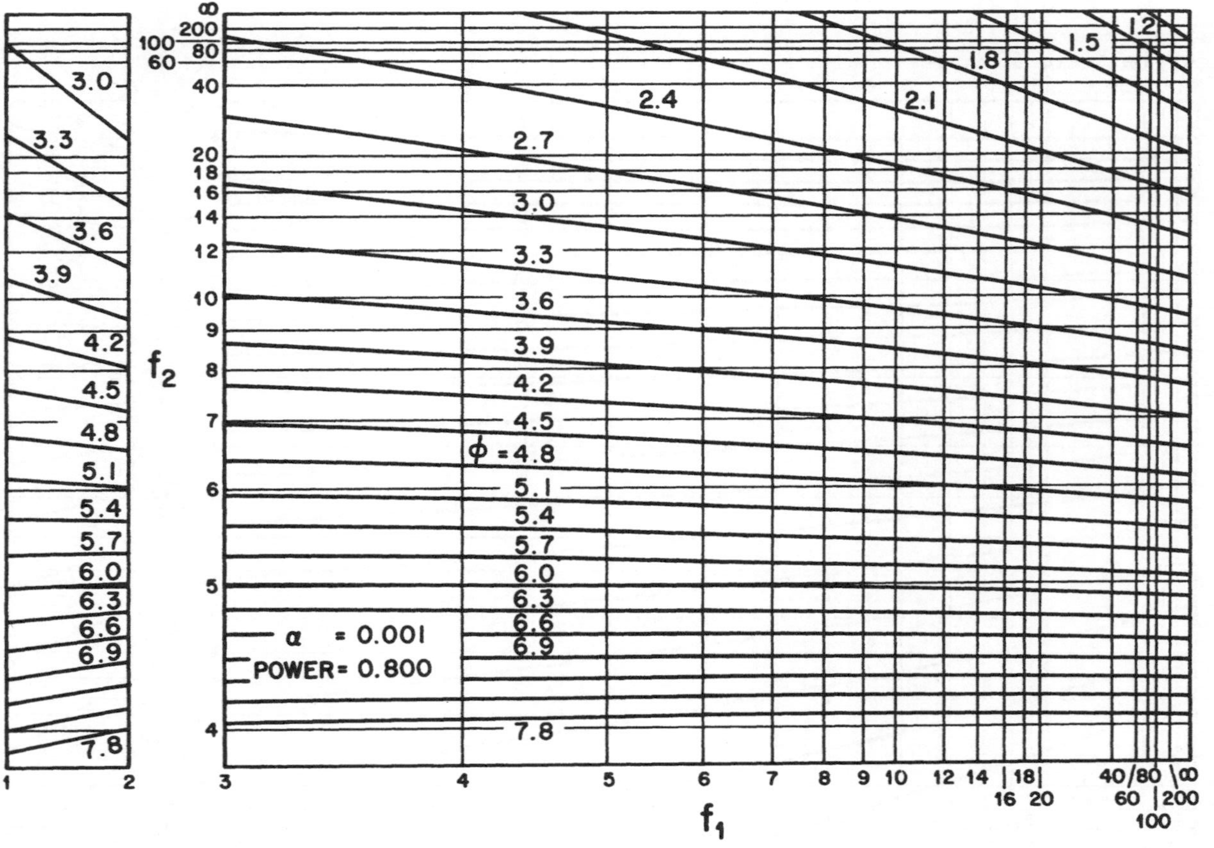

α = 0.001
POWER = 0.800
φ = 4.8
f1
f2

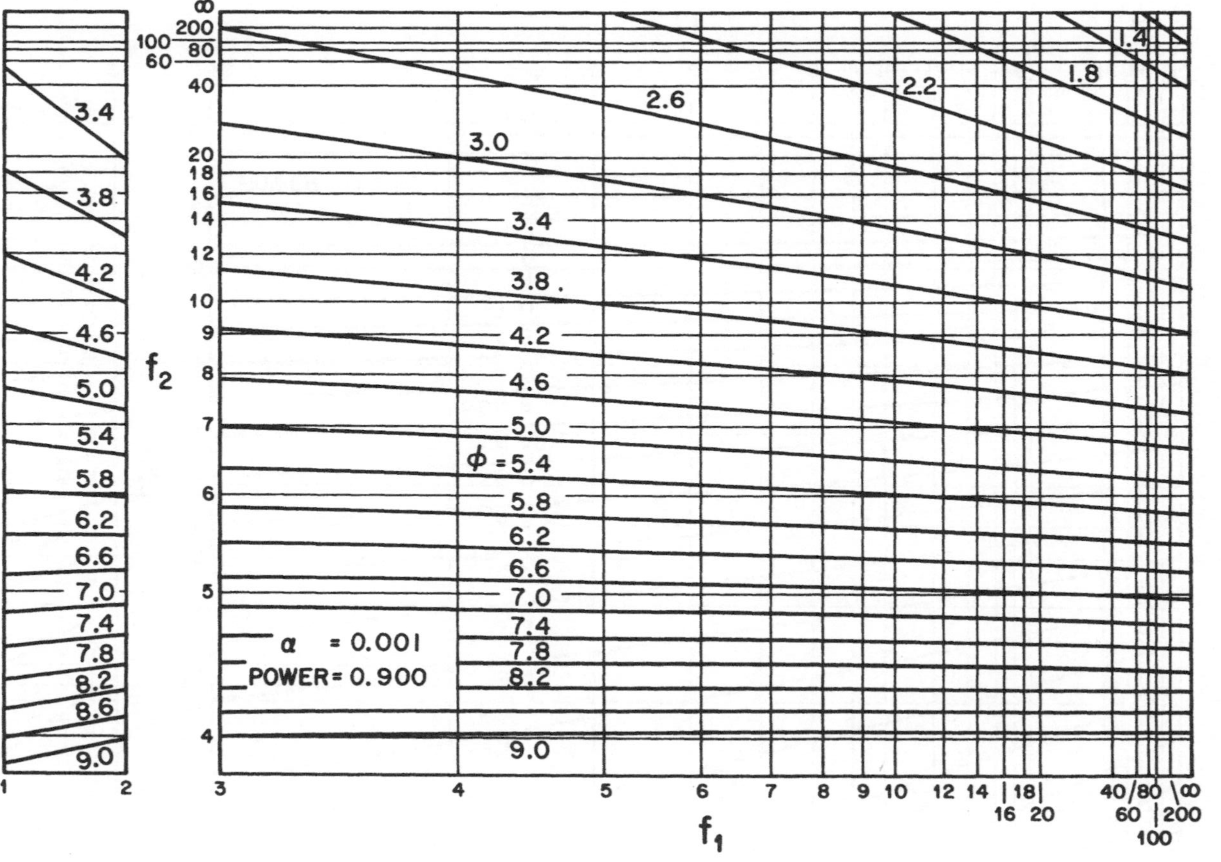
α = 0.001
POWER = 0.900
f1
f2
ϕ = 5.4
1.4
1.8
2.2
2.6
3.0
3.4
3.8
4.2
4.6
5.0
5.4
5.8
6.2
6.6
7.0
7.4
7.8
8.2
8.6
9.0

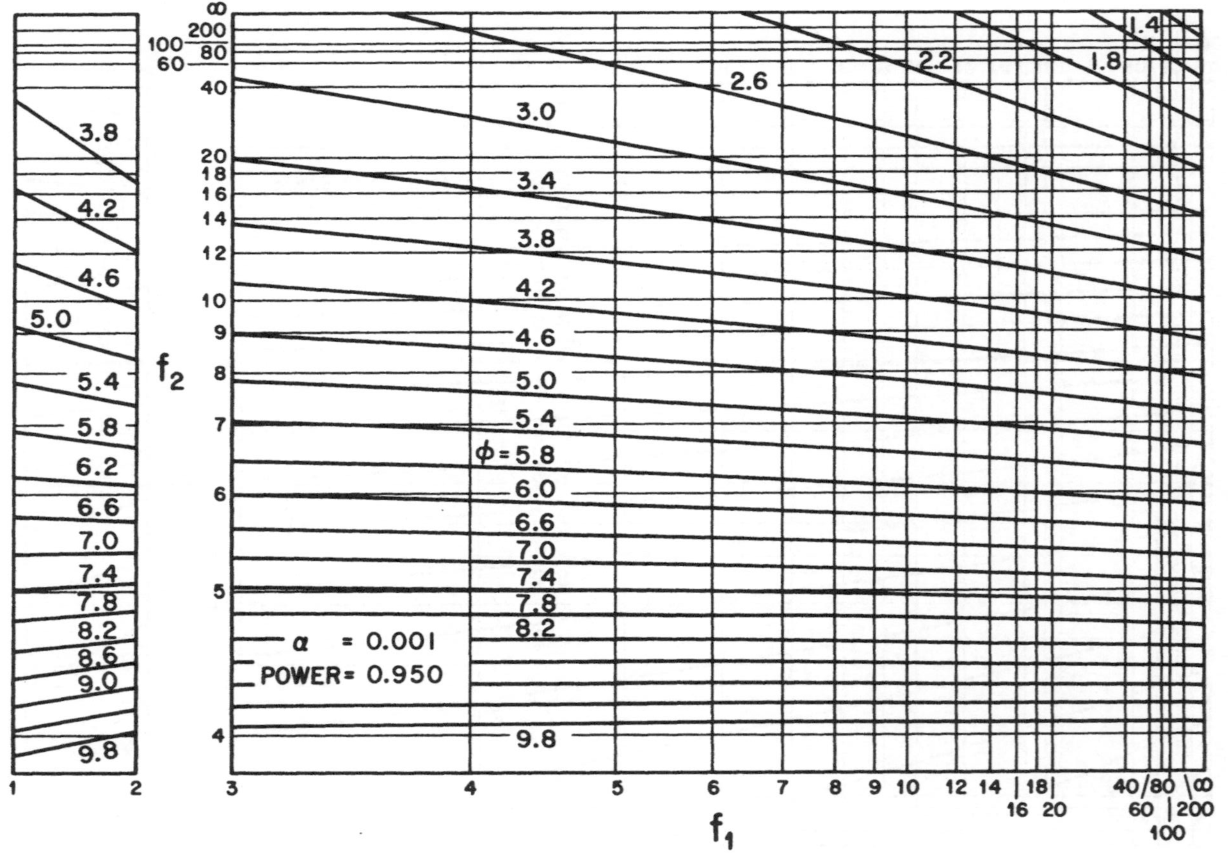
α = 0.001
POWER = 0.950
f_1
f_2
φ = 5.8
1.4
1.8
2.2
2.6
3.0
3.4
3.8
4.2
4.6
5.0
5.4
6.0
6.6
7.0
7.4
7.8
8.2
9.8
5.8
6.2
8.6
9.0

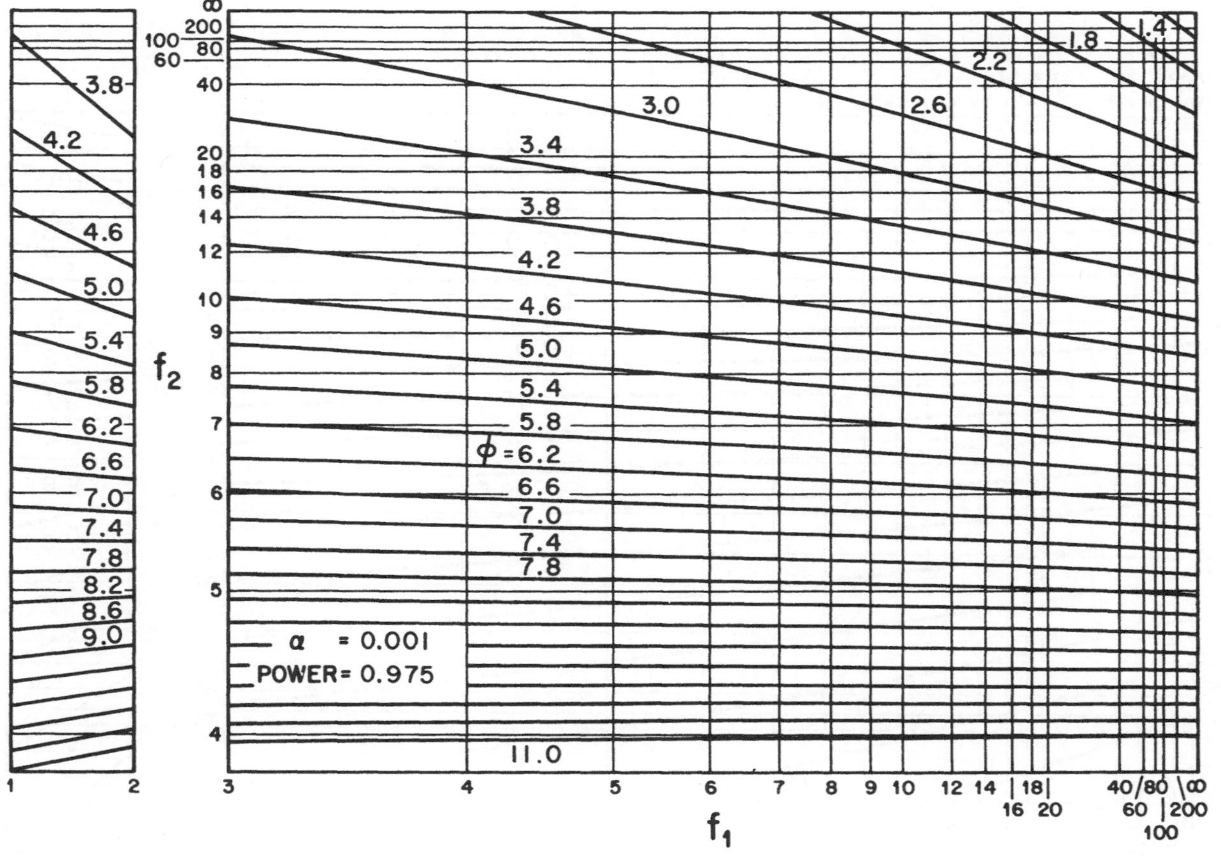
α = 0.001
POWER = 0.975
φ = 6.2
f2
f1

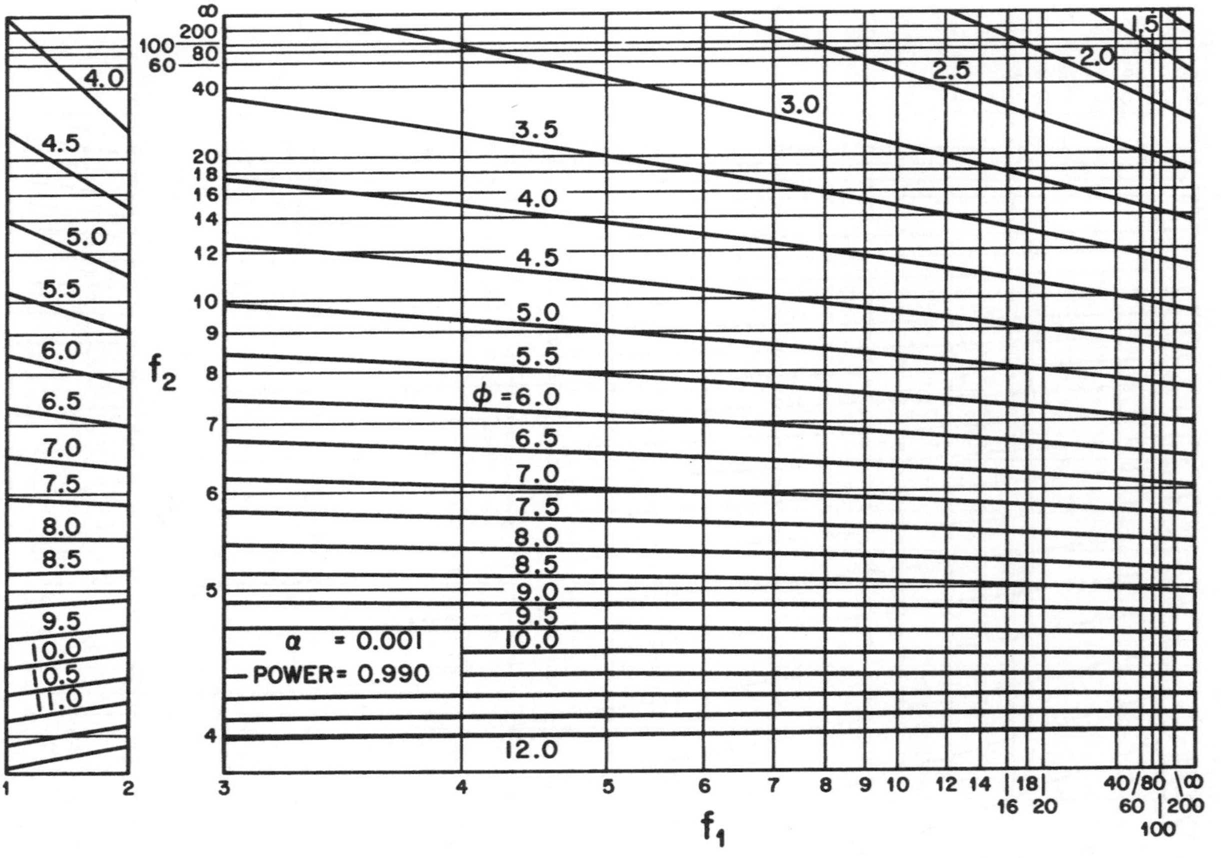

α = 0.001
POWER = 0.990
ϕ = 6.0
1.5
2.0
2.5
3.0
3.5
4.0
4.5
5.0
5.5
6.0
6.5
7.0
7.5
8.0
8.5
9.0
9.5
10.0
10.5
11.0
12.0
f_1
f_2
1 2 3 4 5 6 7 8 9 10 12 14 16 18 20 40 60 80 100 200 ∞
∞ 200 100 80 60 40 20 18 16 14 12 10 9 8 7 6 5 4

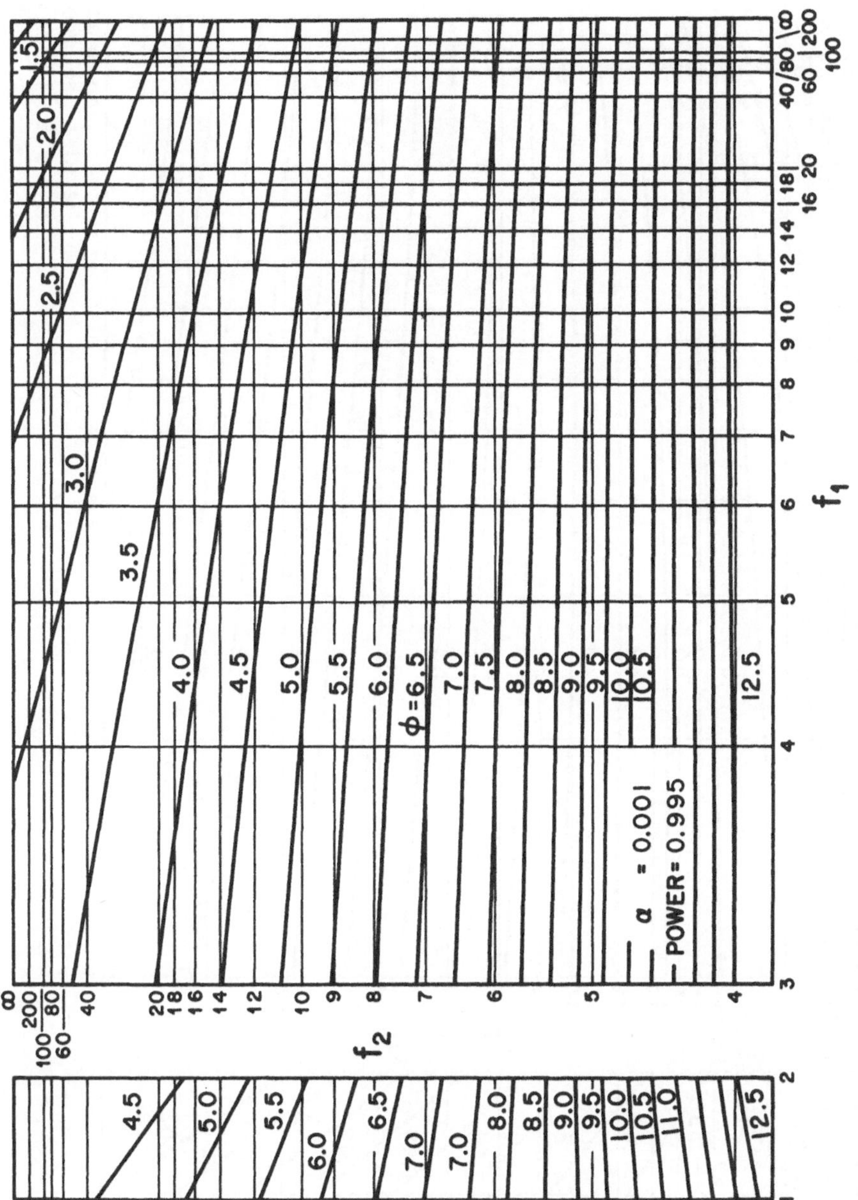
α = 0.001
POWER = 0.995
ϕ = 6.5
f_1
f_2

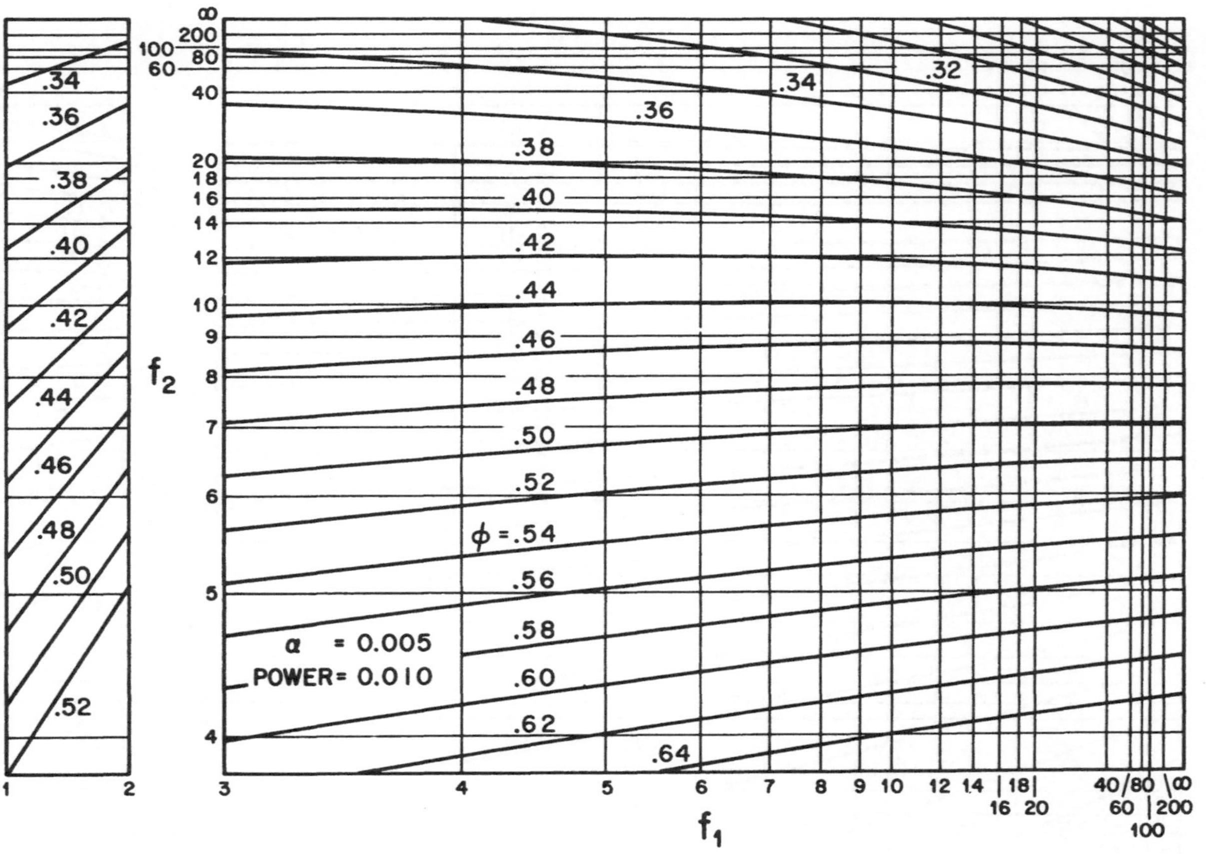
α = 0.005
POWER = 0.010
φ = .54
f_1
f_2
.32
.34
.36
.38
.40
.42
.44
.46
.48
.50
.52
.56
.58
.60
.62
.64

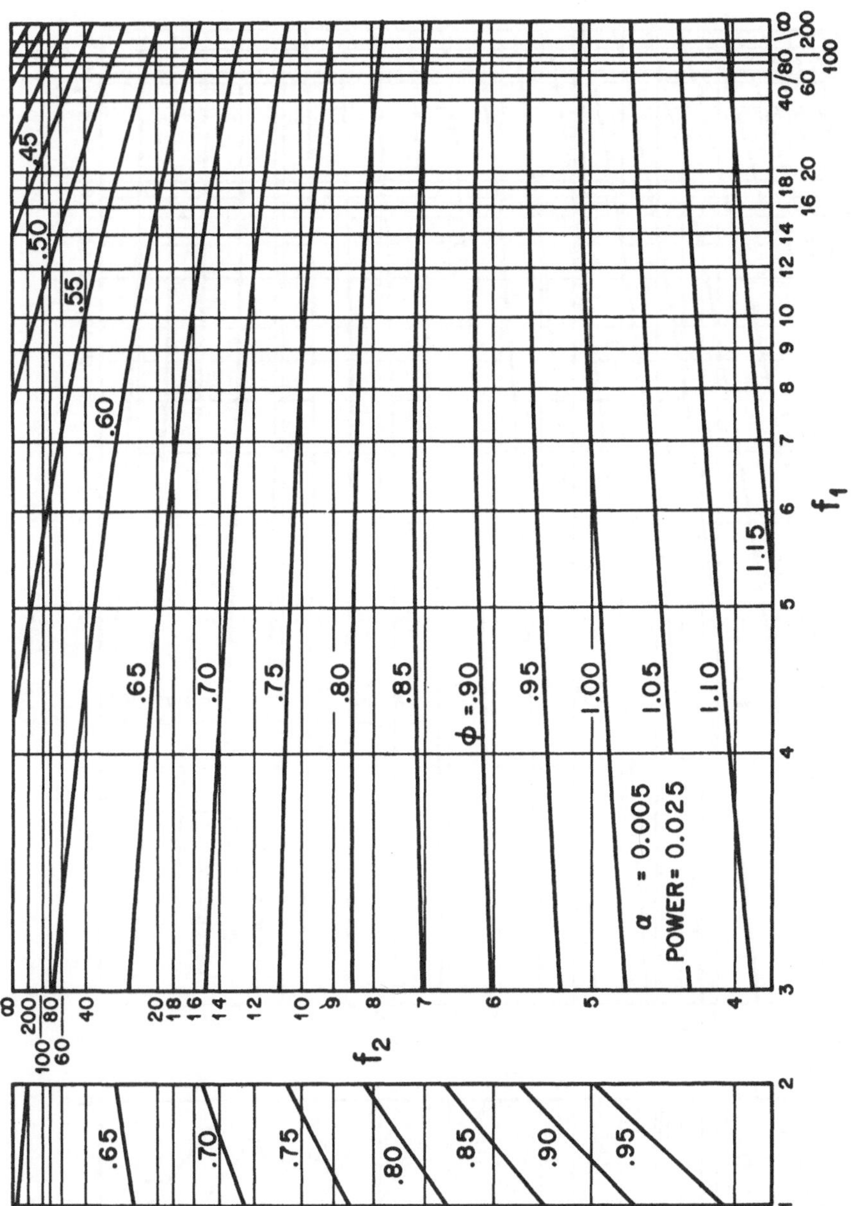

α = 0.005
POWER = 0.025
φ = .90
.45
.50
.55
.60
.65
.70
.75
.80
.85
.95
1.00
1.05
1.10
1.15
f_1
f_2

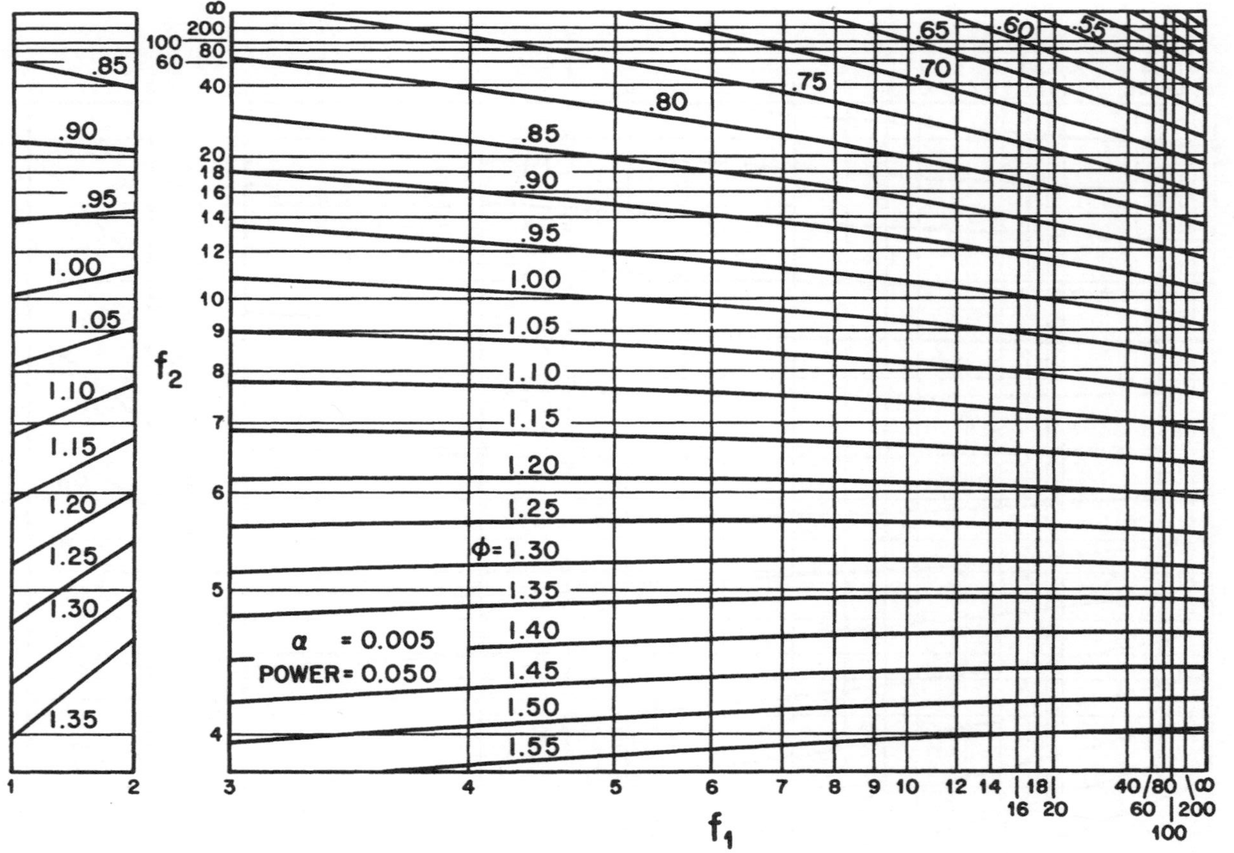

α = 0.005
POWER = 0.050
φ = 1.30
.55
.60
.65
.70
.75
.80
.85
.90
.95
1.00
1.05
1.10
1.15
1.20
1.25
1.30
1.35
1.40
1.45
1.50
1.55
f1
f2

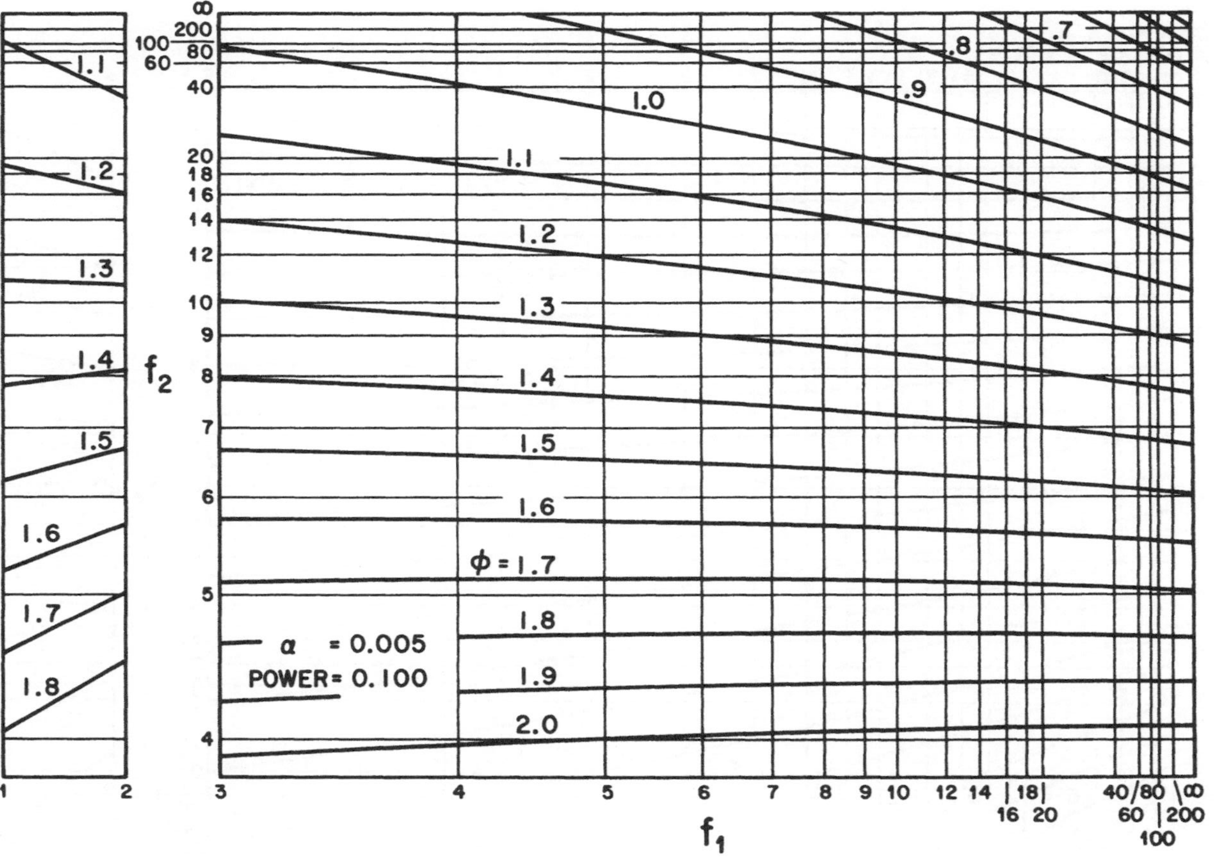
α = 0.005
POWER = 0.100
φ = 1.7
.7
.8
.9
1.0
1.1
1.2
1.3
1.4
1.5
1.6
1.8
1.9
2.0
f_1
f_2

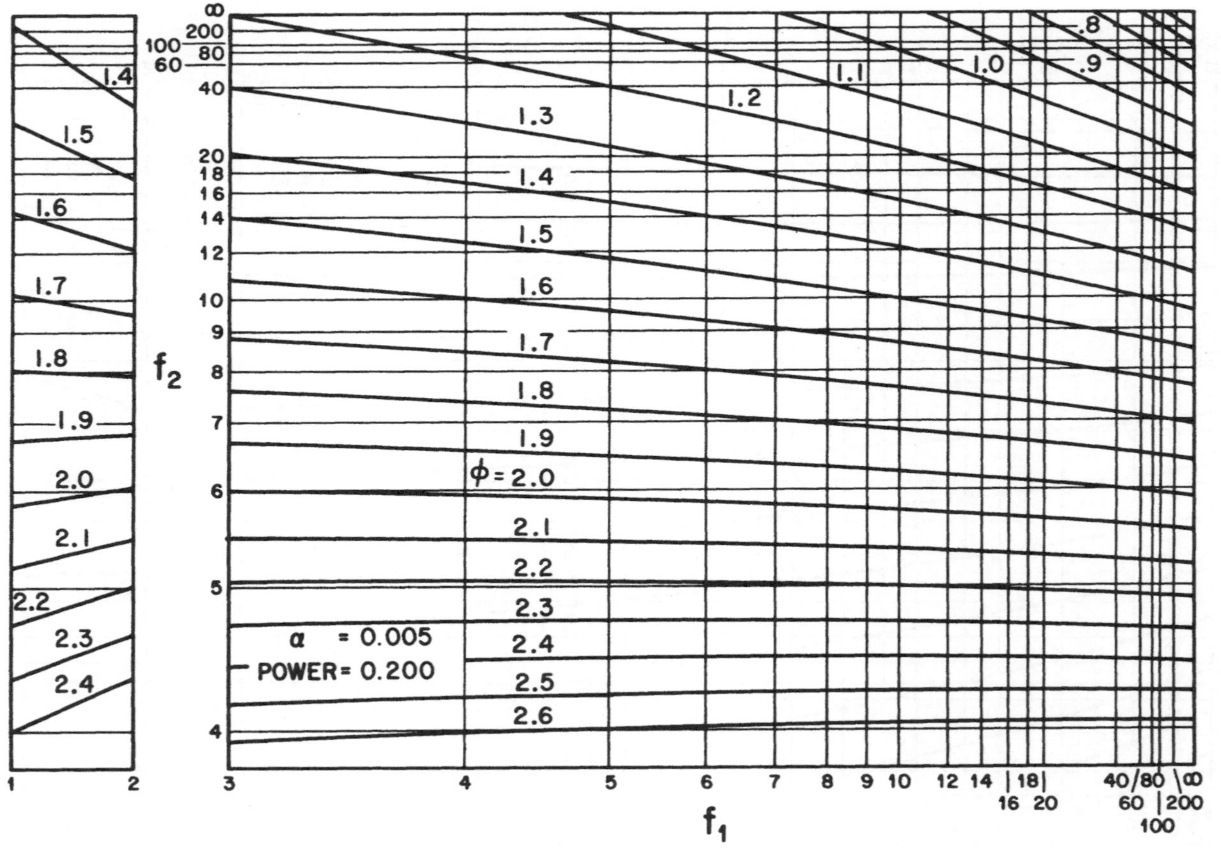

α = 0.005
POWER = 0.200
φ = 2.0
f_1
f_2

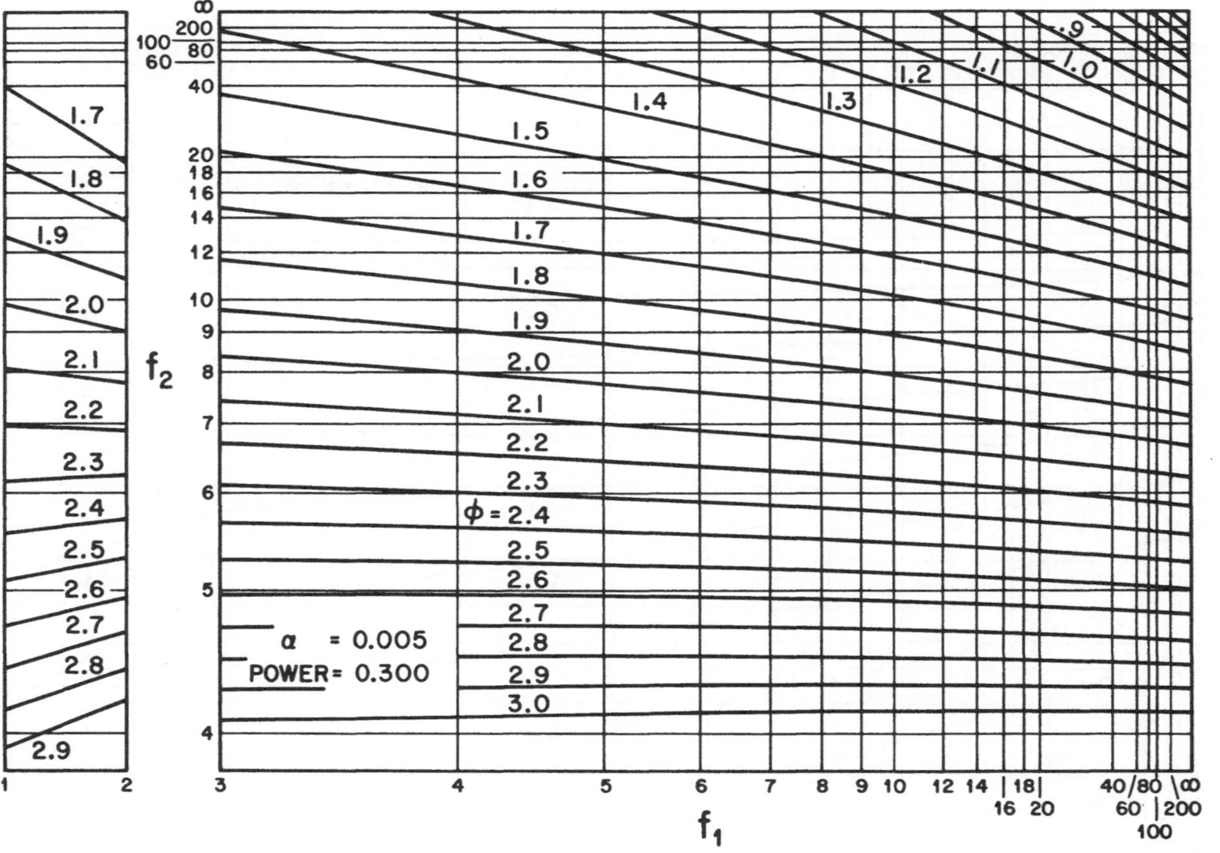
α = 0.005
POWER = 0.300
φ = 2.4
f_1
f_2

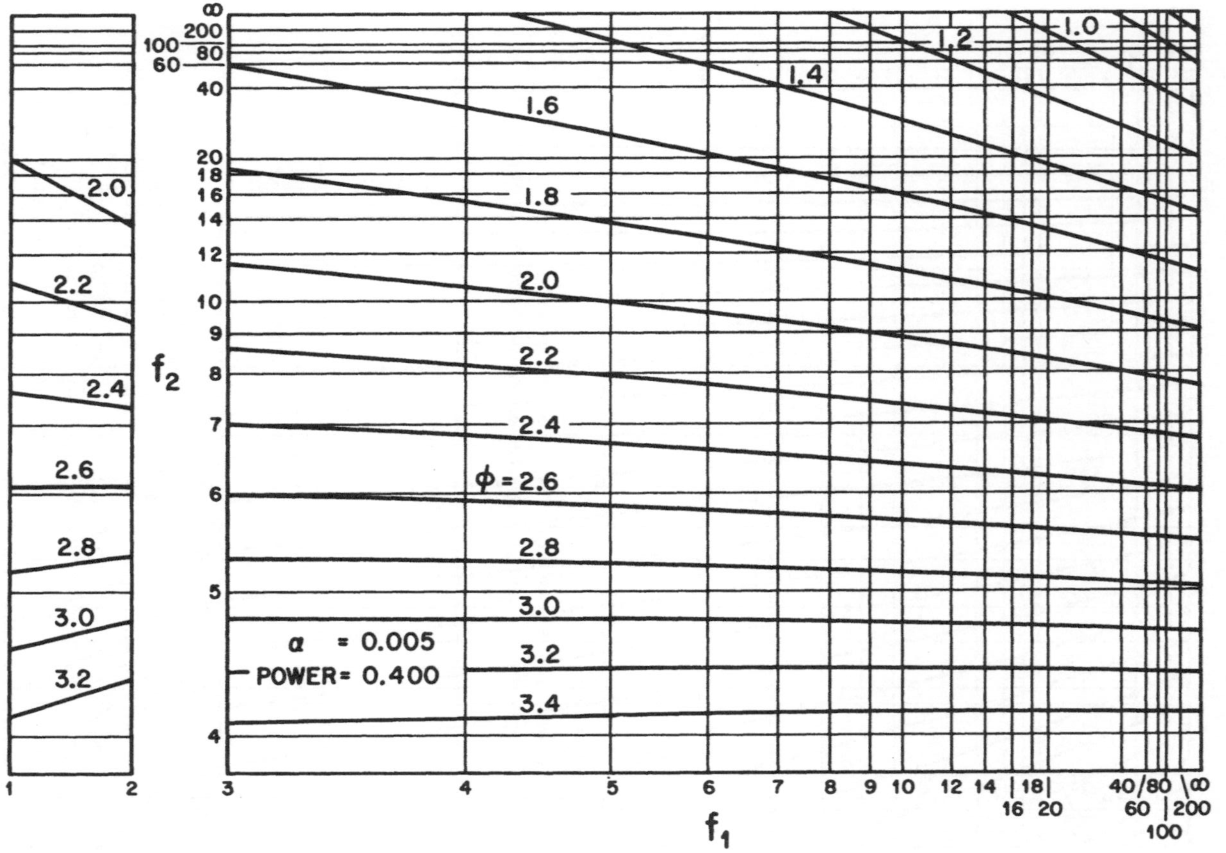

α = 0.005
POWER = 0.400
φ = 2.6
1.0
1.2
1.4
1.6
1.8
2.0
2.2
2.4
2.6
2.8
3.0
3.2
3.4
f_1
f_2

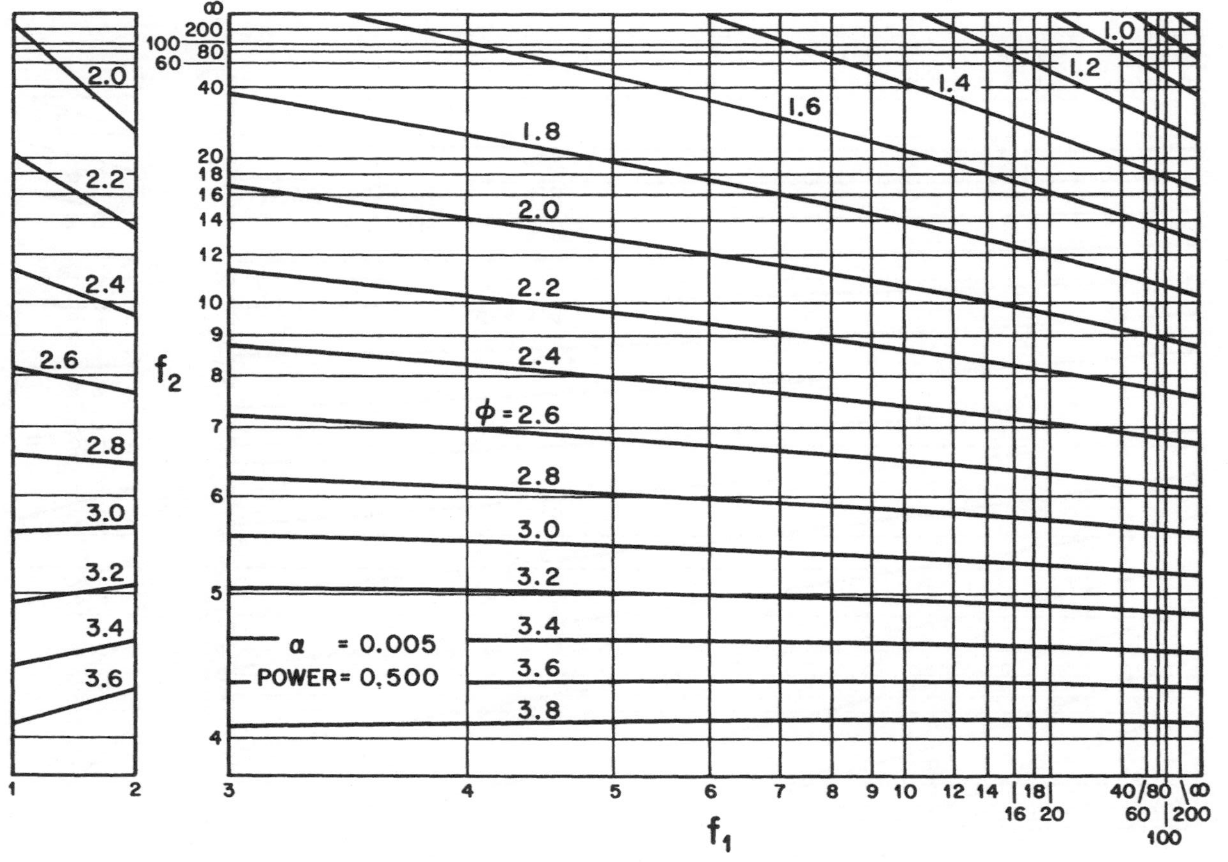
α = 0.005
POWER = 0,500
φ = 2.6
1.0
1.2
1.4
1.6
1.8
2.0
2.2
2.4
2.8
3.0
3.2
3.4
3.6
3.8
f_1
f_2

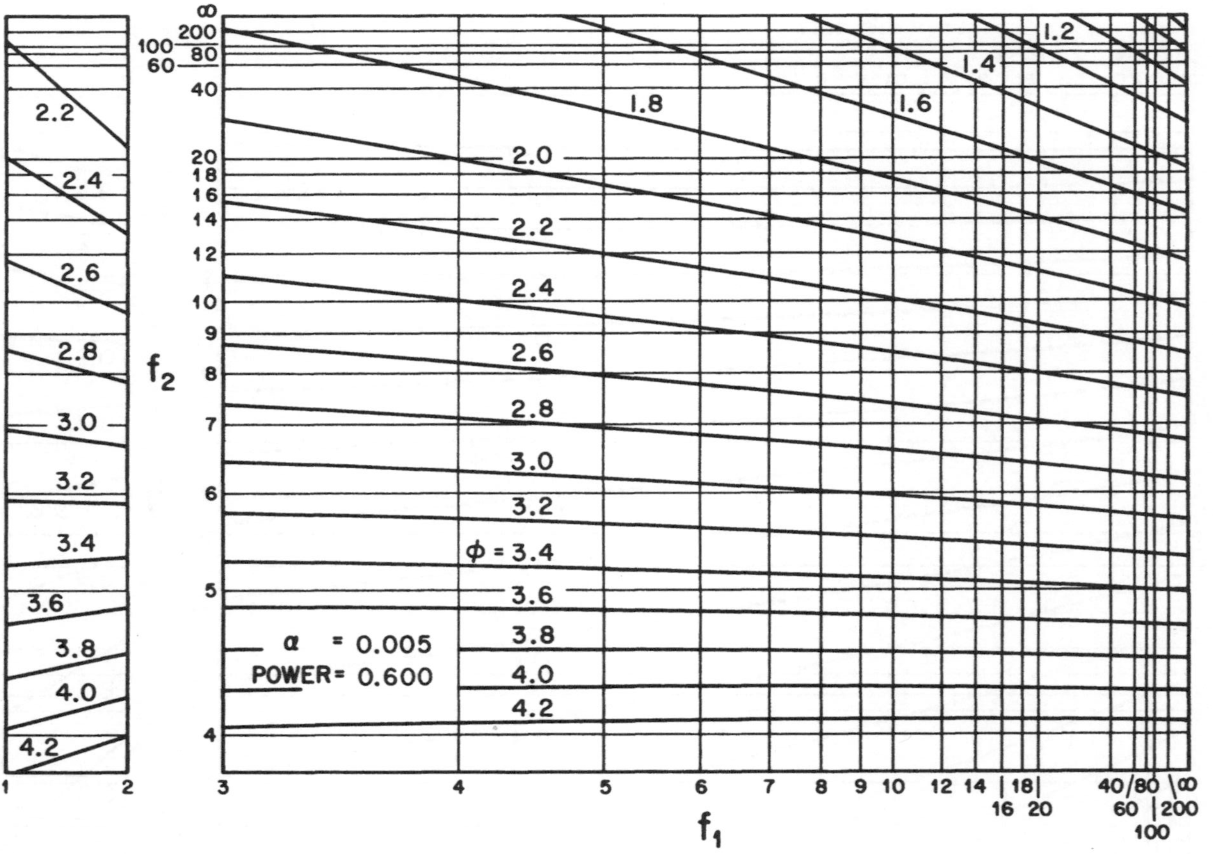

α = 0.005
POWER = 0.600
φ = 3.4
f_1
f_2

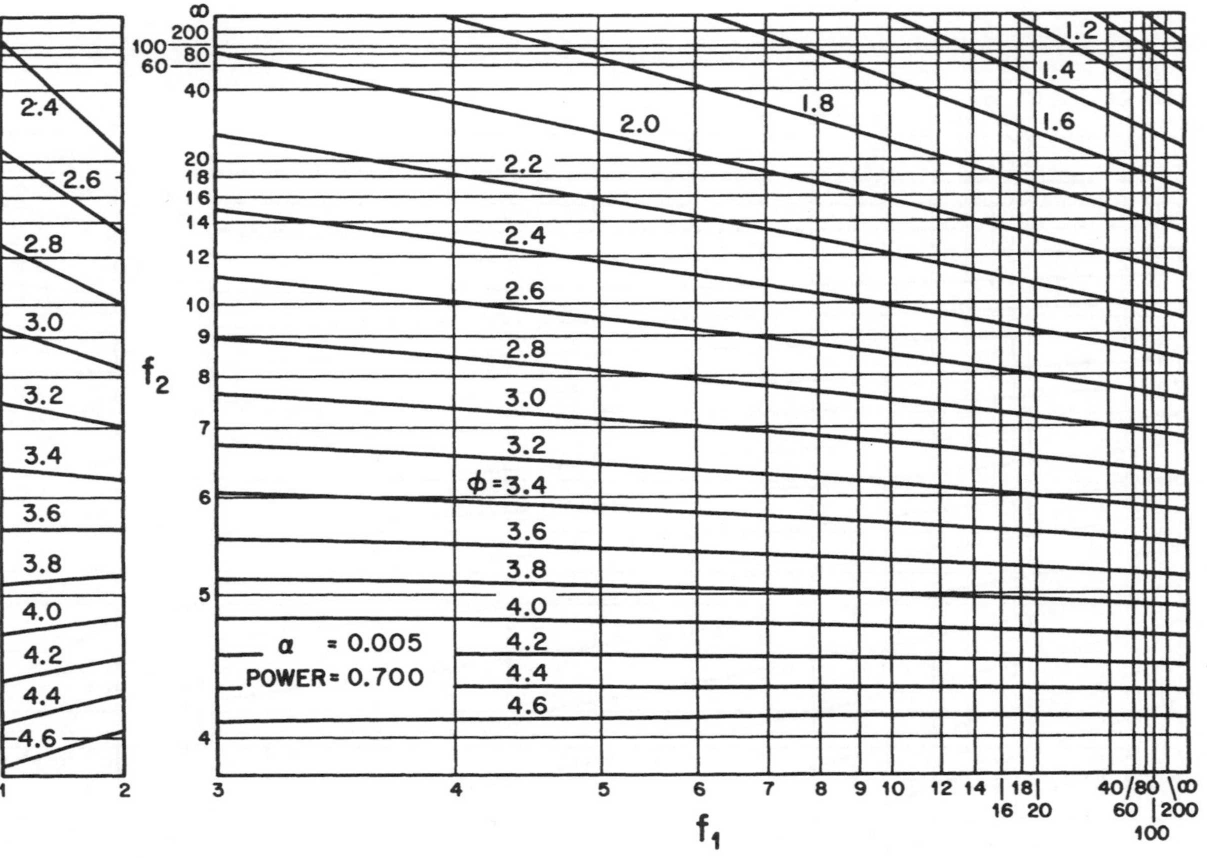
α = 0.005
POWER = 0.700
f_1
f_2
φ = 3.4
1.2
1.4
1.6
1.8
2.0
2.2
2.4
2.6
2.8
3.0
3.2
3.6
3.8
4.0
4.2
4.4
4.6

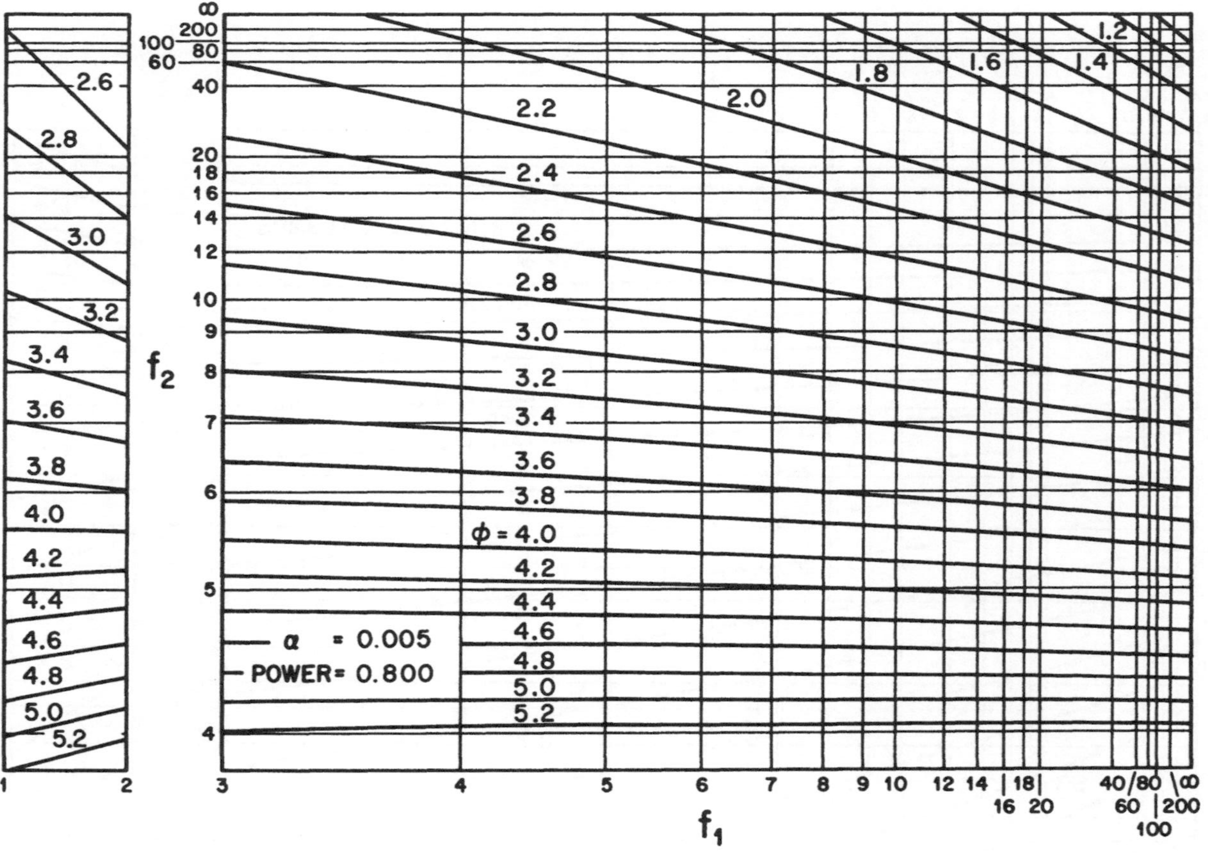

α = 0.005
POWER = 0.800
φ = 4.0
f_1
f_2

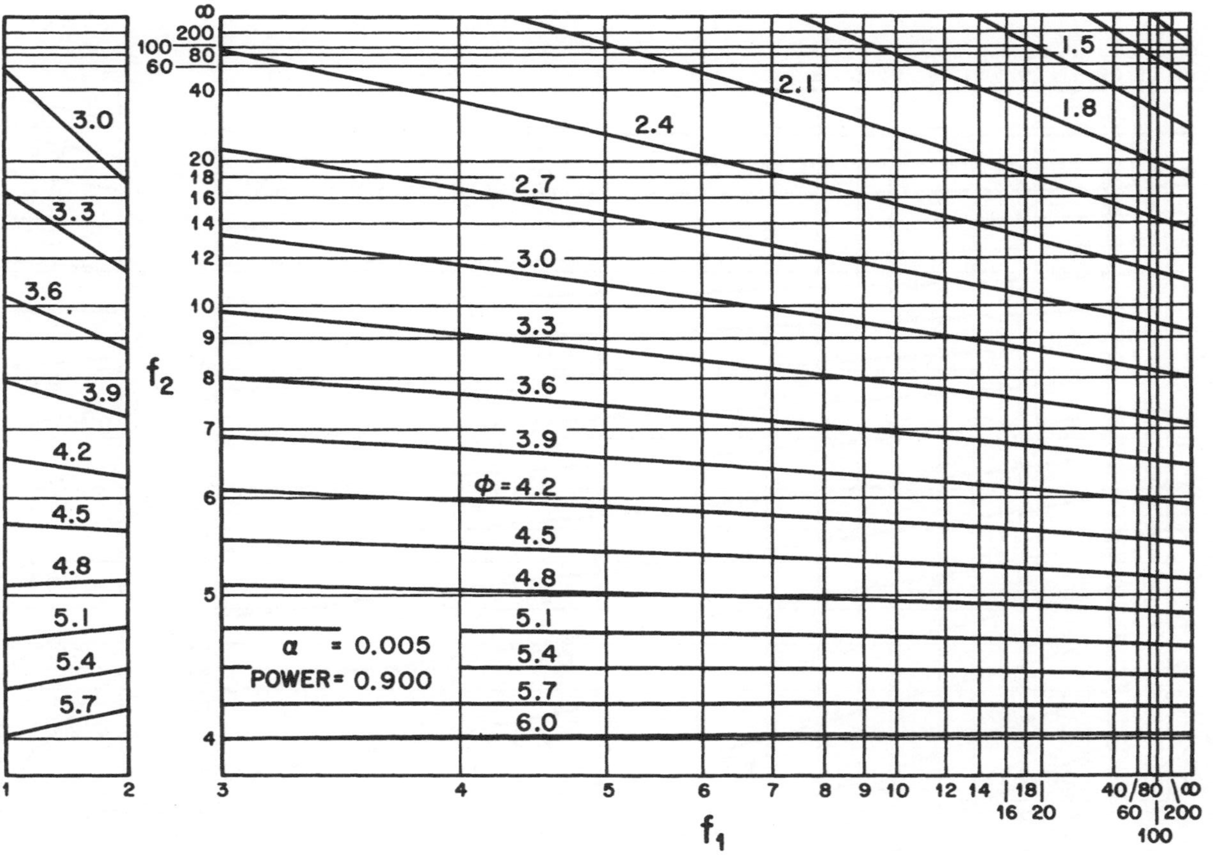

α = 0.005
POWER = 0.900
φ = 4.2
1.5
1.8
2.1
2.4
2.7
3.0
3.3
3.6
3.9
4.5
4.8
5.1
5.4
5.7
6.0
3.0
3.3
3.6
3.9
4.2
4.5
4.8
5.1
5.4
5.7
f_1
f_2

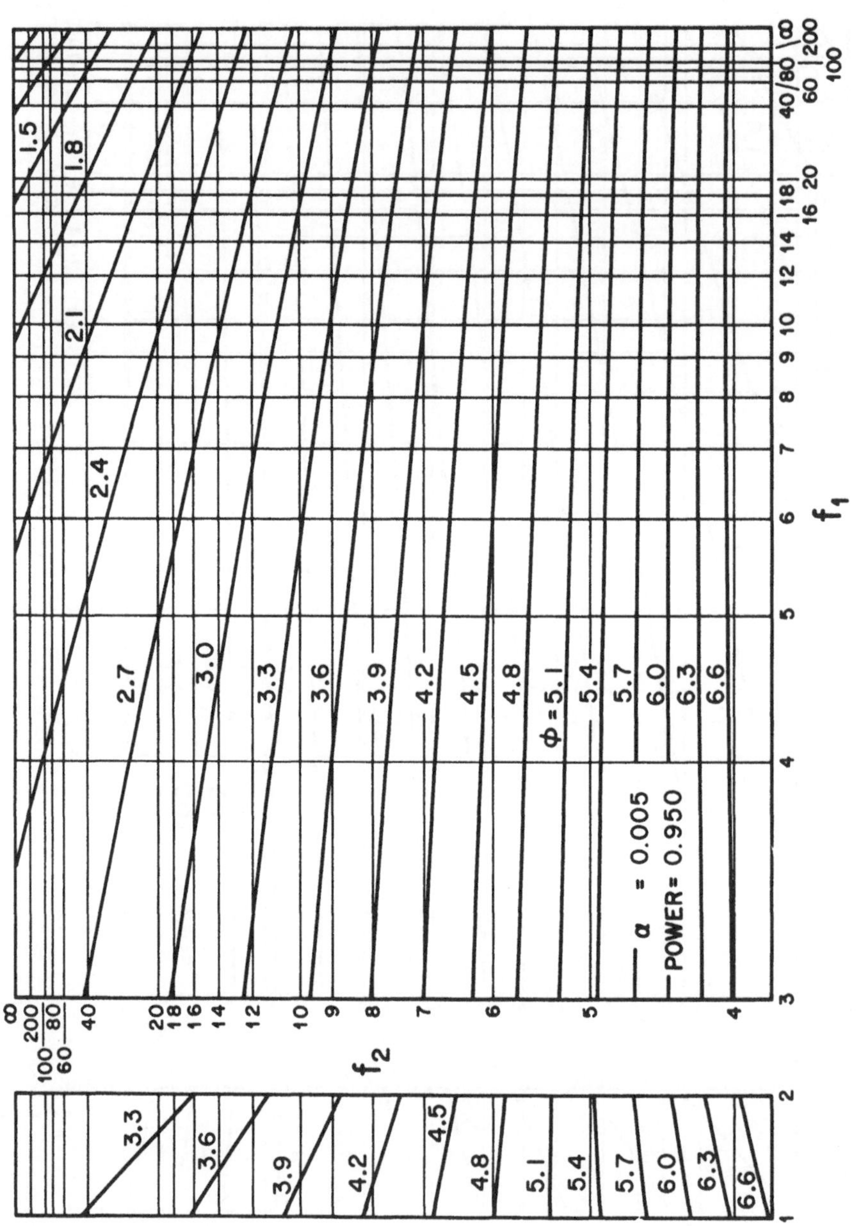

α = 0.005
POWER = 0.950
φ = 5.1
f_1
f_2

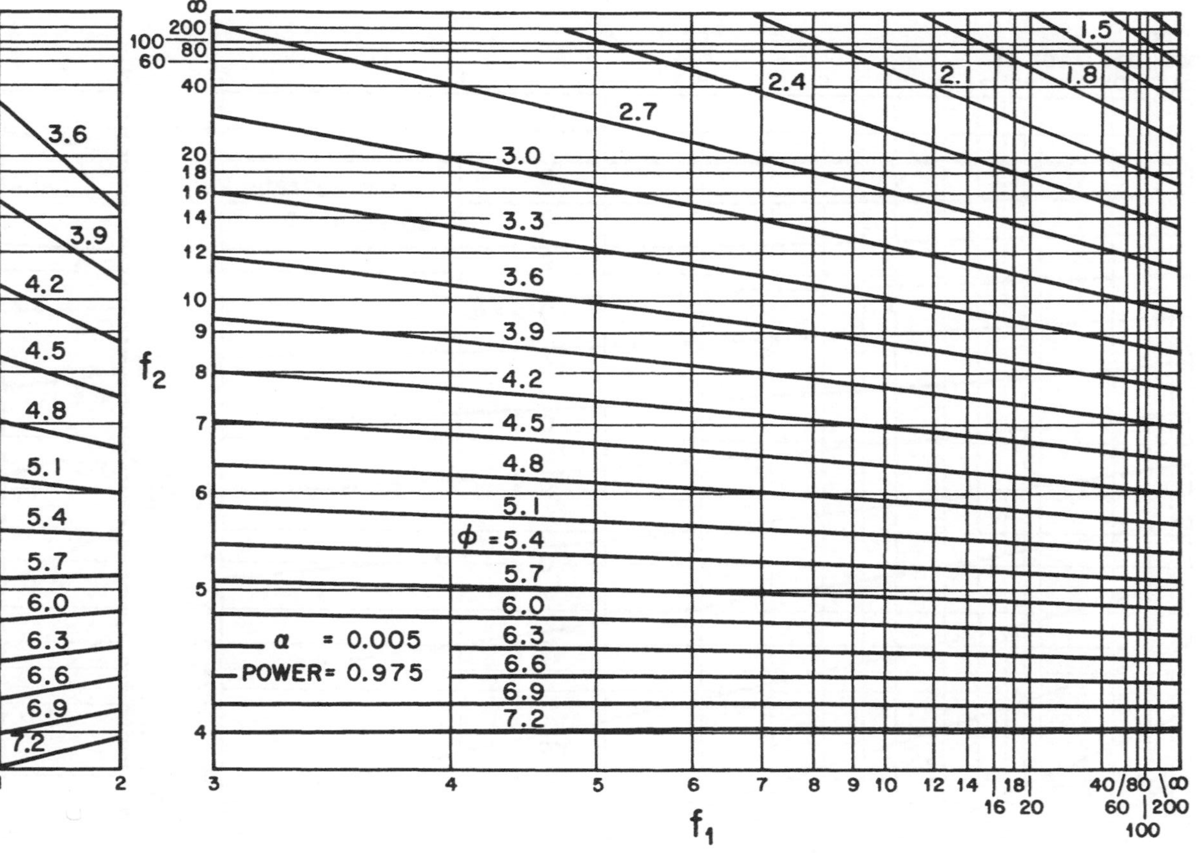
α = 0.005
POWER= 0.975
φ = 5.4
1.5
1.8
2.1
2.4
2.7
3.0
3.3
3.6
3.9
4.2
4.5
4.8
5.1
5.7
6.0
6.3
6.6
6.9
7.2
f1
1 2 3 4 5 6 7 8 9 10 12 14 16 18 20 40 60 80 100 200 ∞
f2
∞ 200 100 80 60 40 20 18 16 14 12 10 9 8 7 6 5 4

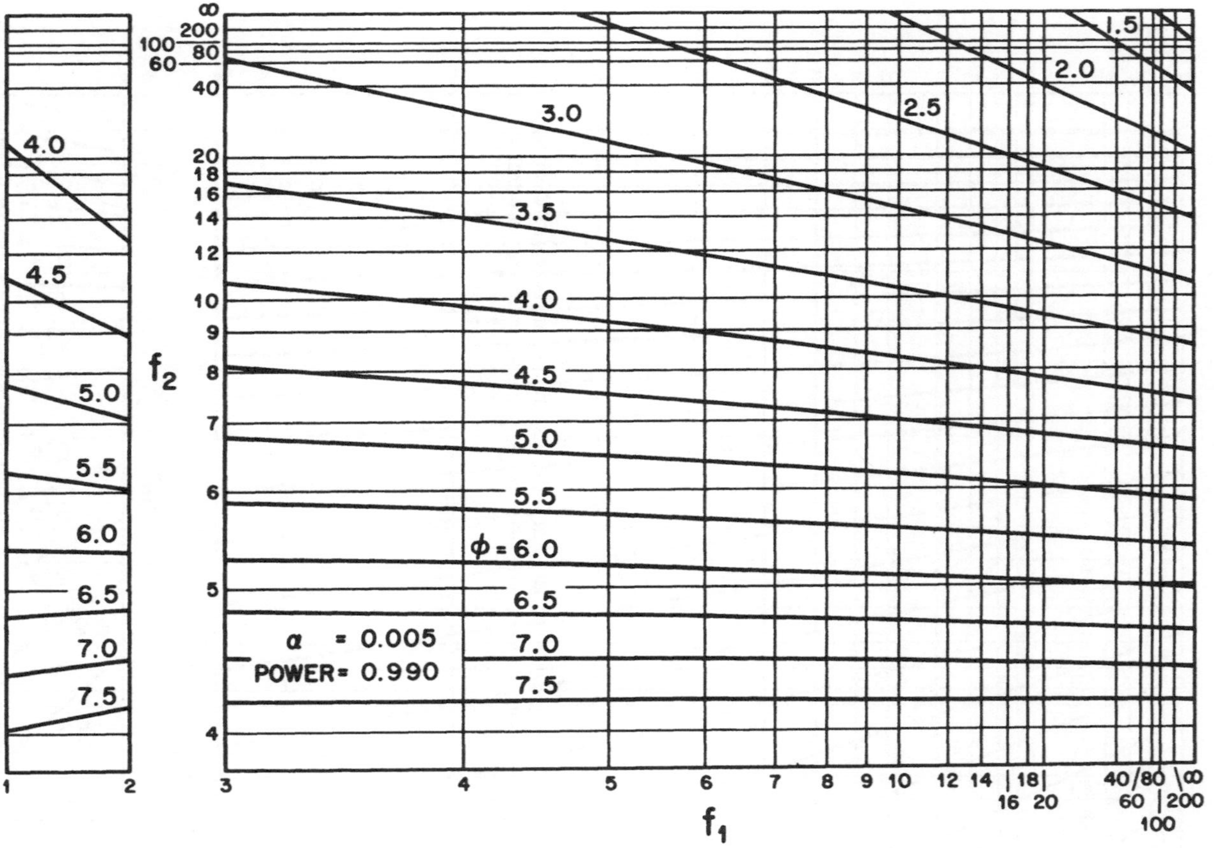

α = 0.005
POWER = 0.990
φ = 6.0
1.5
2.0
2.5
3.0
3.5
4.0
4.5
5.0
5.5
6.5
7.0
7.5
f1
f2

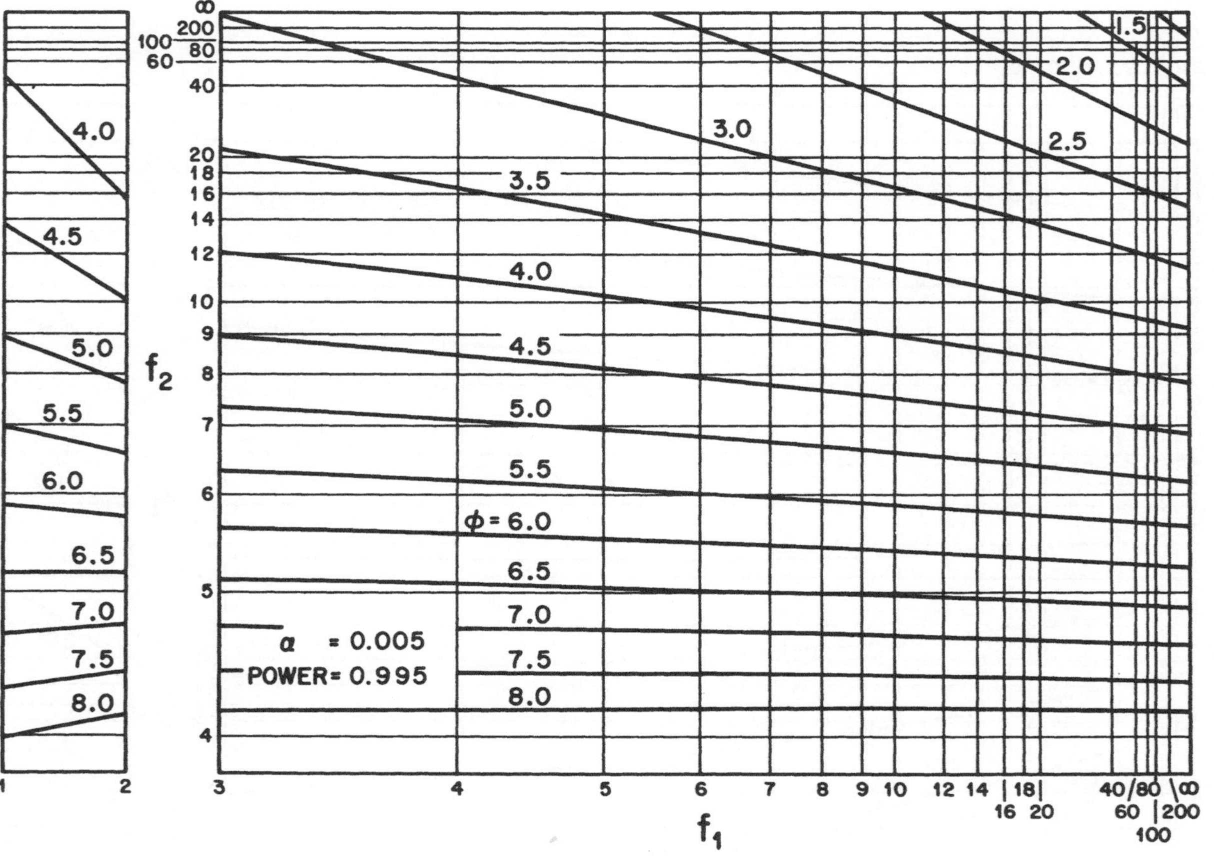

α = 0.005
POWER = 0.995
ϕ = 6.0
1.5
2.0
2.5
3.0
3.5
4.0
4.5
5.0
5.5
6.0
6.5
7.0
7.5
8.0
f_1
f_2

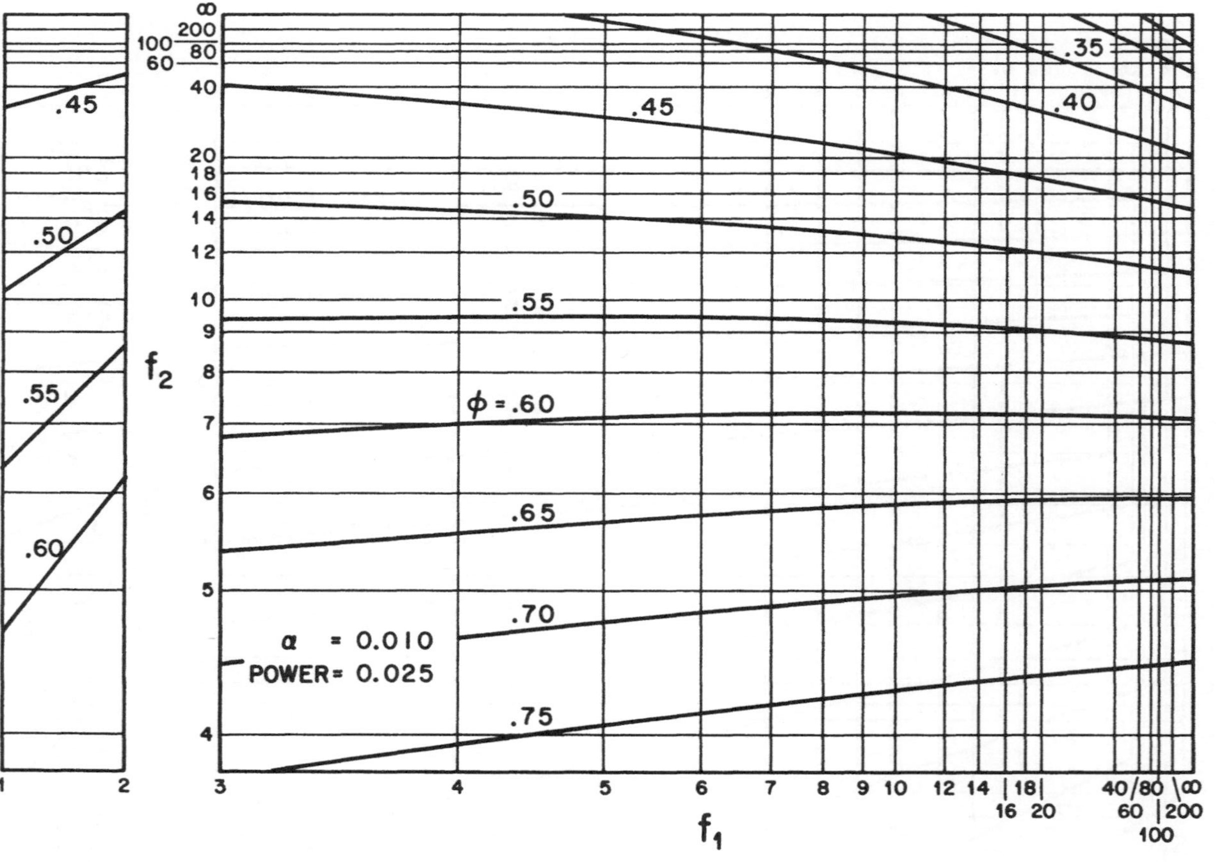

α = 0.010
POWER = 0.025
φ = .60
.35
.40
.45
.50
.55
.65
.70
.75
.45
.50
.55
.60
f_1
f_2
1
2
3
4
5
6
7
8
9
10
12
14
16
18
20
40
60
80
100
200
∞

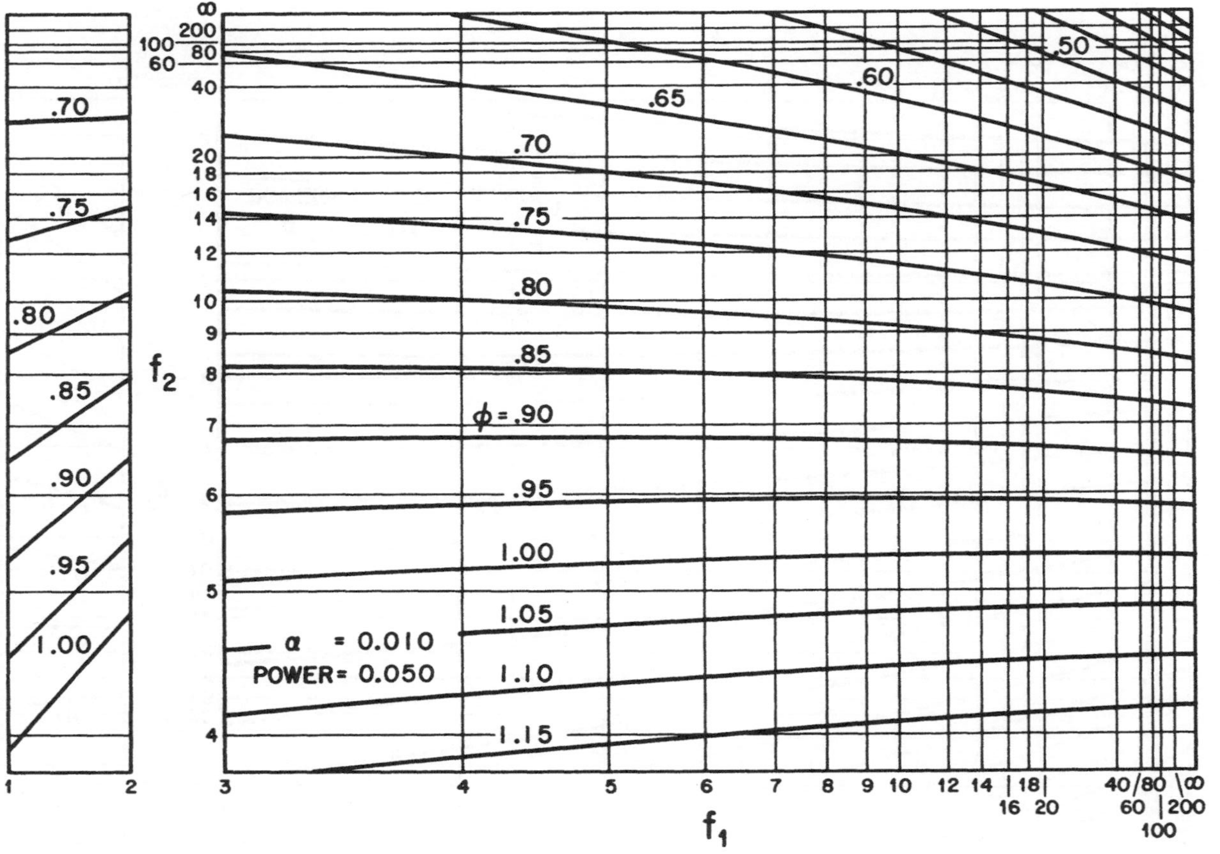

α = 0.010
POWER = 0.050
φ = .90
.50
.60
.65
.70
.75
.80
.85
.95
1.00
1.05
1.10
1.15
f_1
f_2

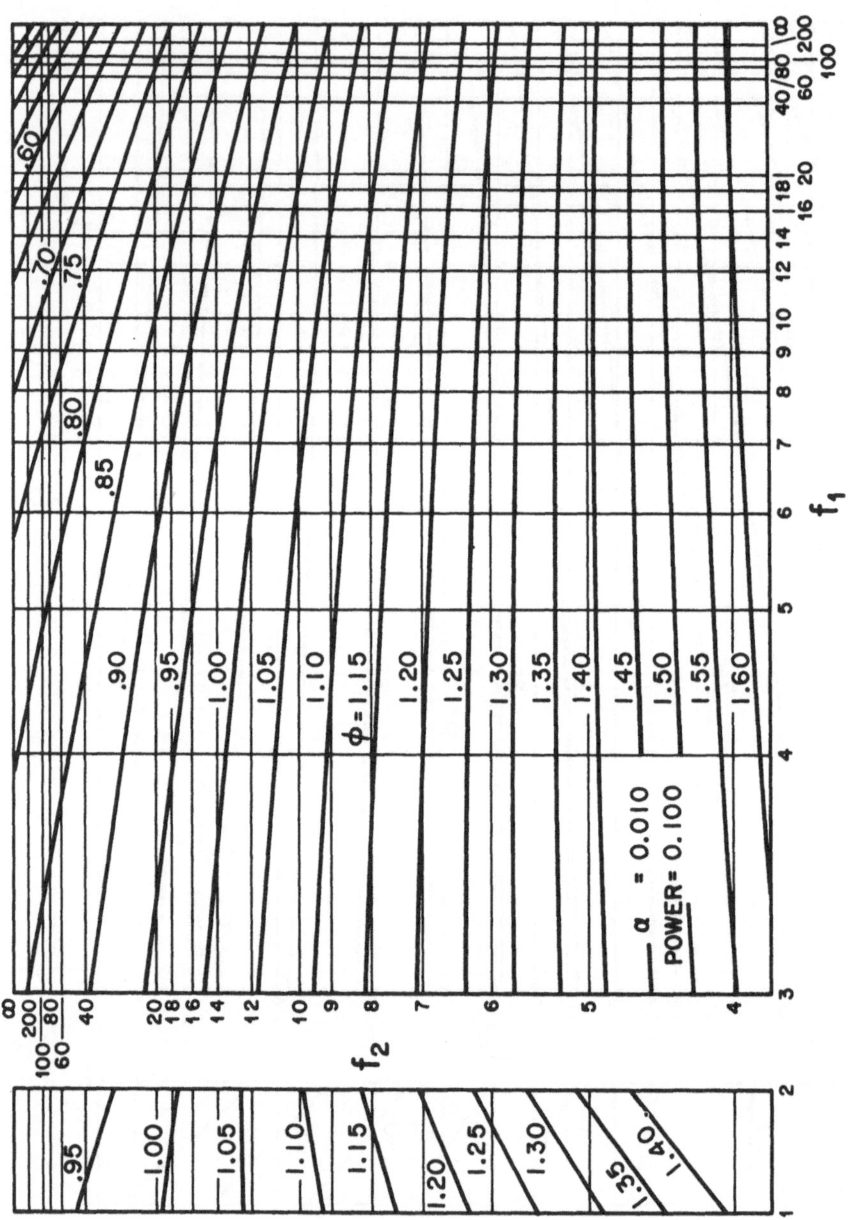

α = 0.010
POWER = 0.100
ϕ = 1.15
f_1
f_2

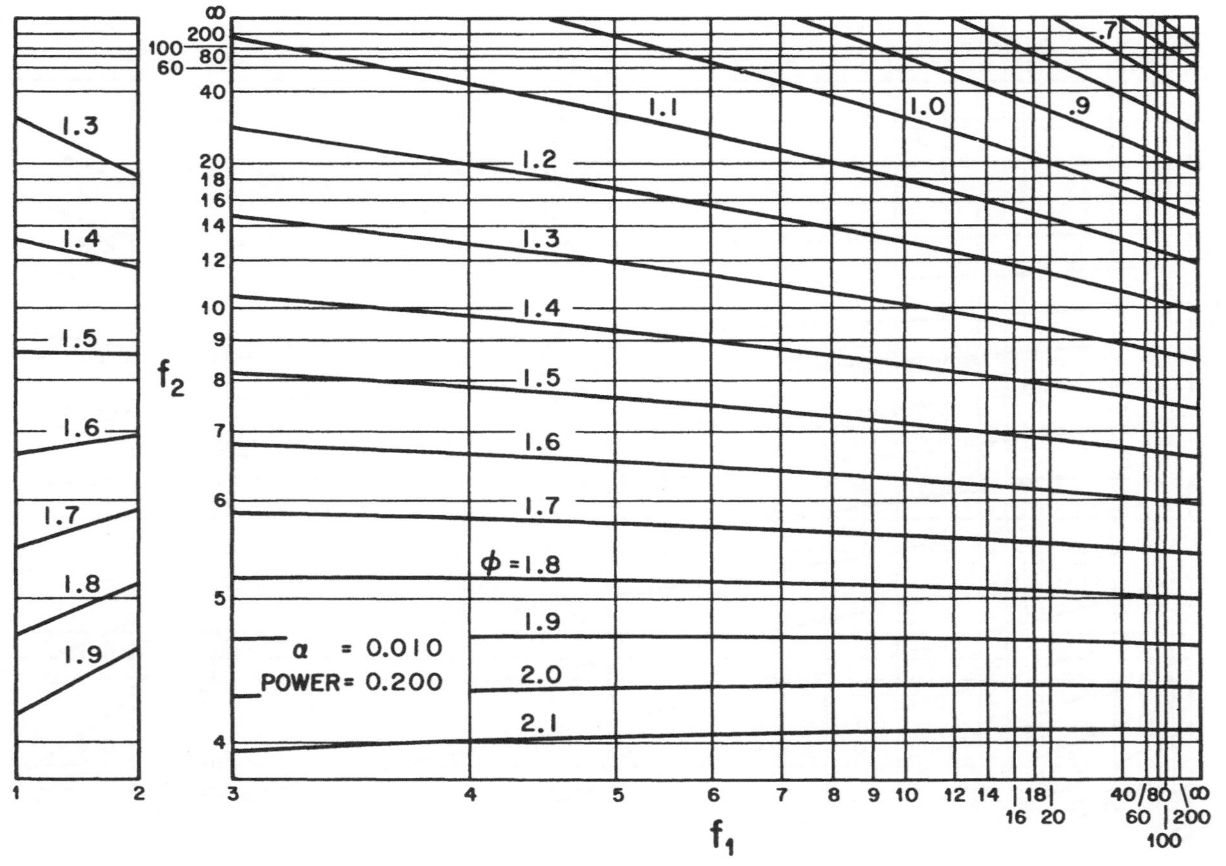

α = 0.010
POWER = 0.200
φ = 1.8
.7
.9
1.0
1.1
1.2
1.3
1.4
1.5
1.6
1.7
1.9
2.0
2.1
1.8
f1
f2

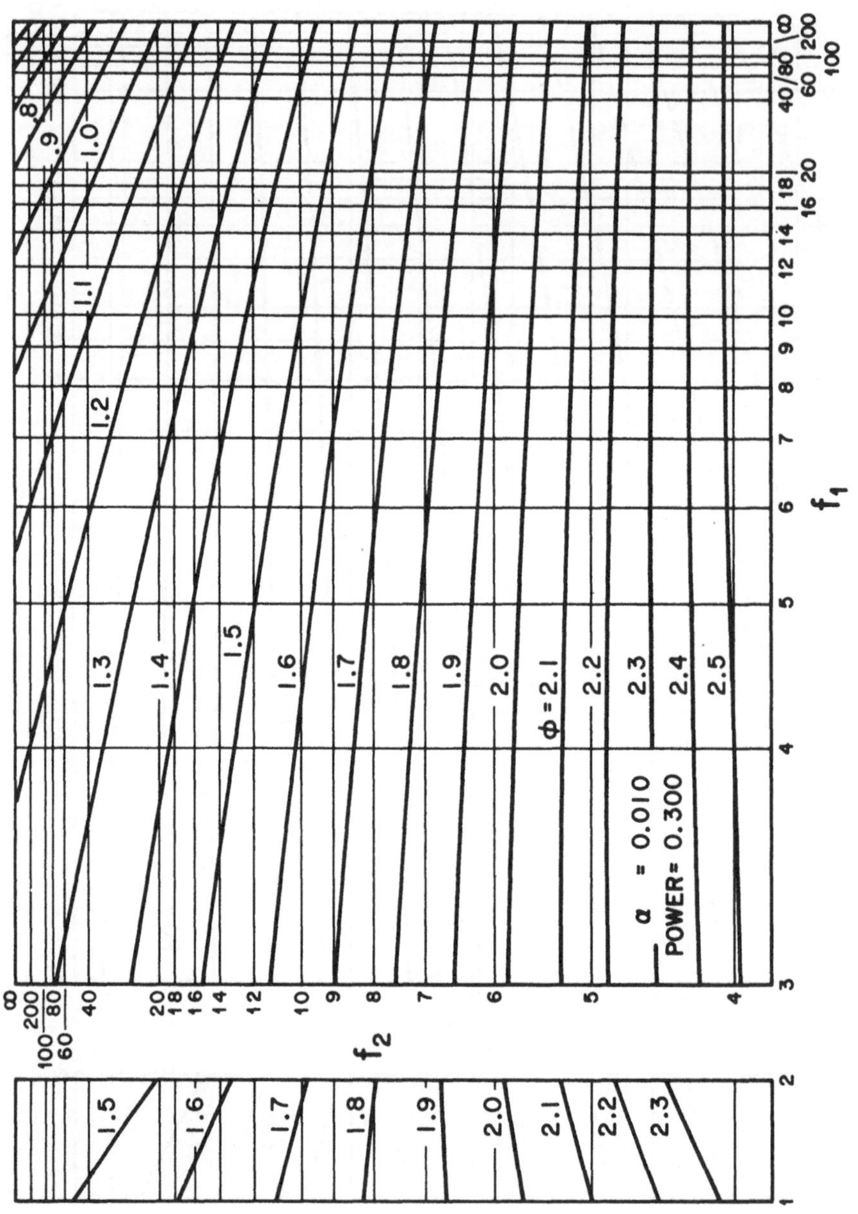

α = 0.010
POWER = 0.300
φ = 2.1
f1
f2

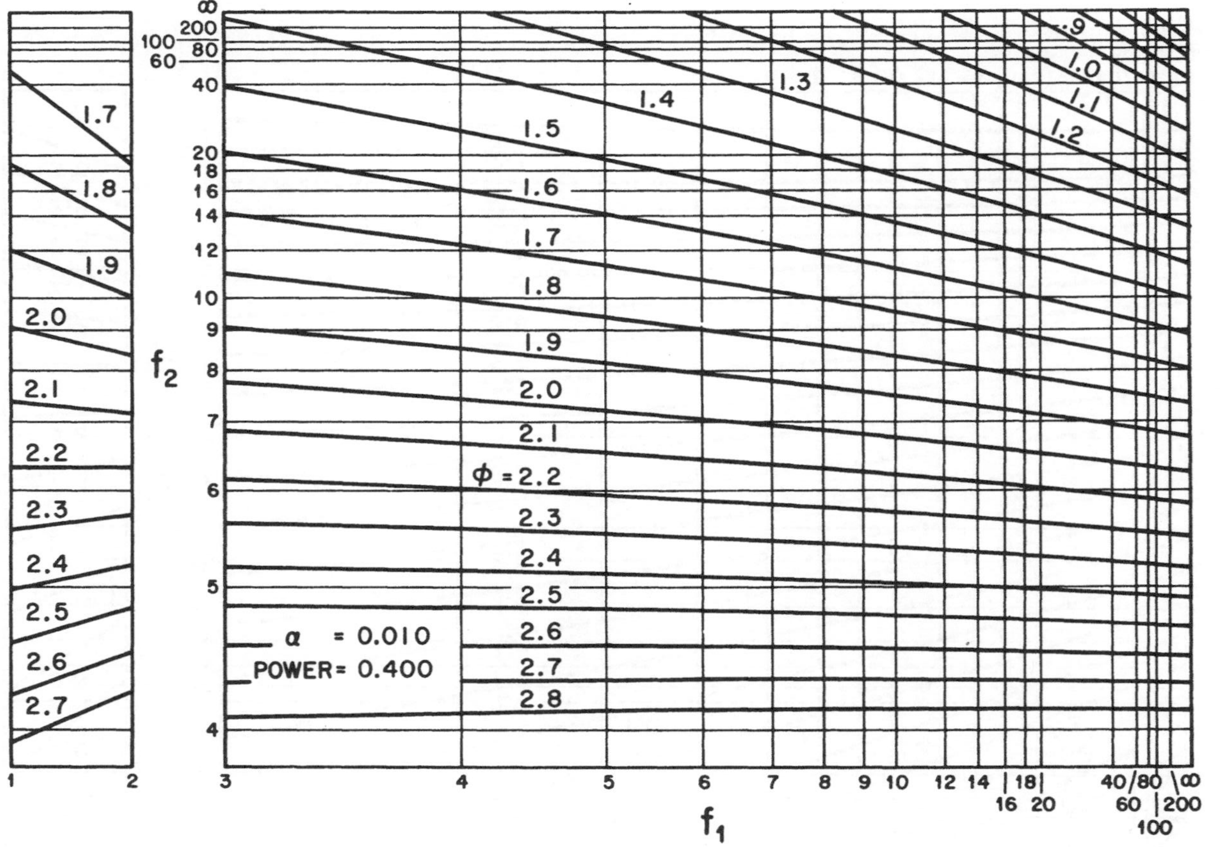

α = 0.010
POWER = 0.400
φ = 2.2
f_1
f_2

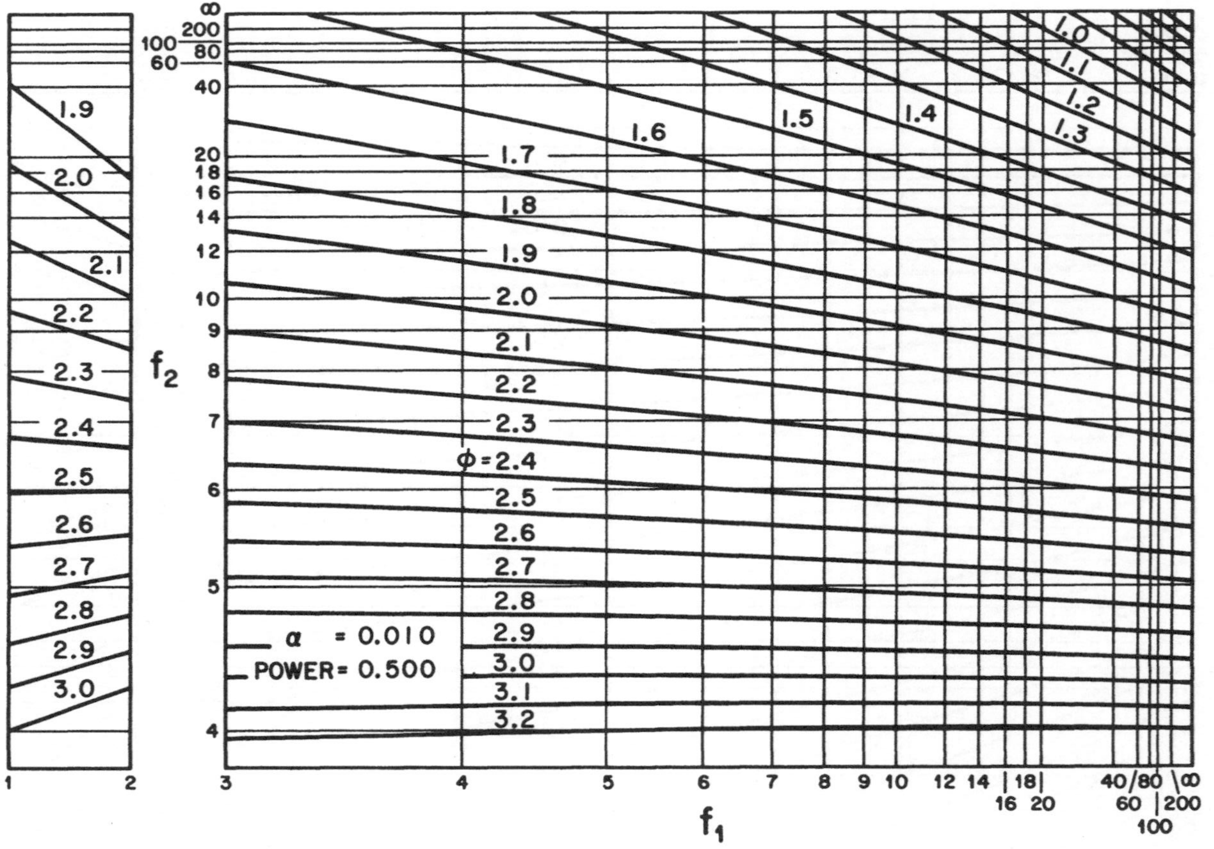
α = 0.010
POWER = 0.500
φ = 2.4
f_1
f_2

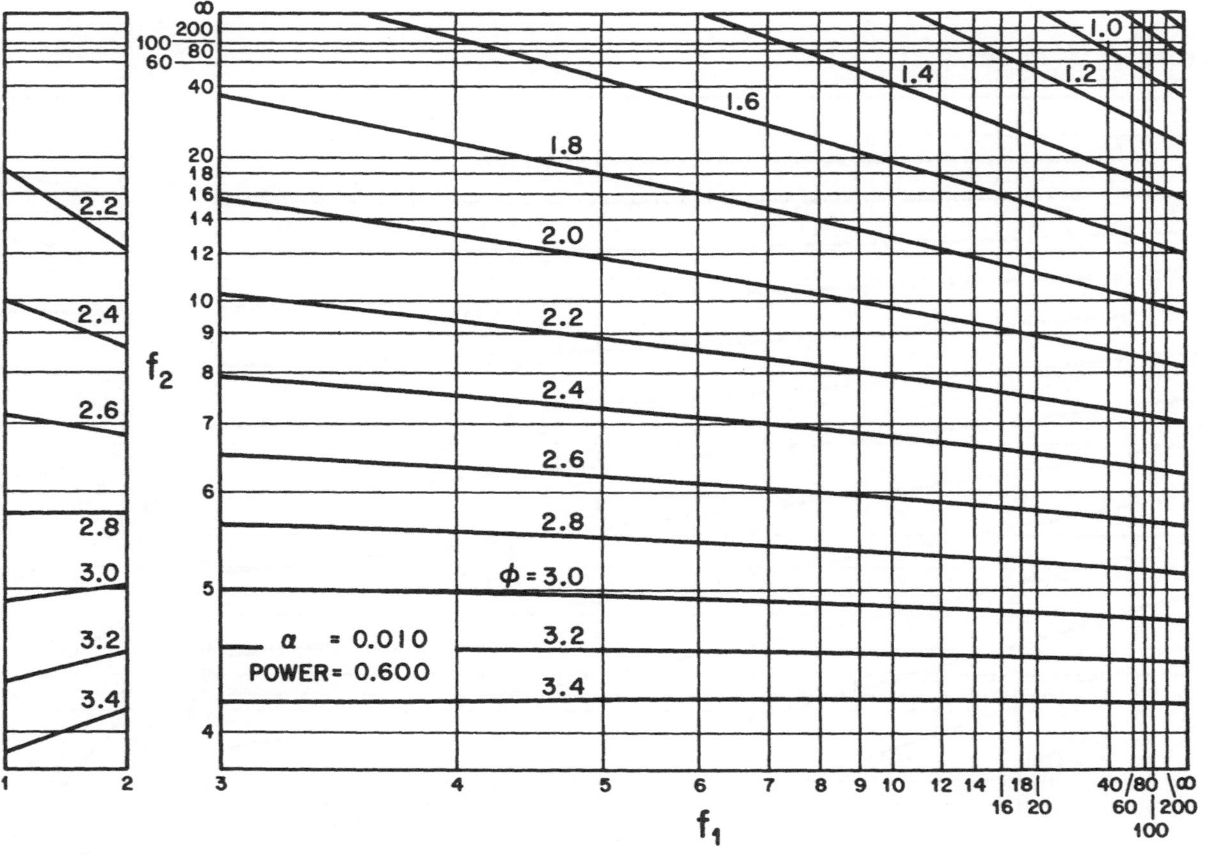

α = 0.010
POWER = 0.600
φ = 3.0
1.0
1.2
1.4
1.6
1.8
2.0
2.2
2.4
2.6
2.8
3.2
3.4
2.2
2.4
2.6
2.8
3.0
3.2
3.4
f_1
f_2
1 2 3 4 5 6 7 8 9 10 12 14 16 18 20 40 60 80 100 200 ∞
∞ 200 100 80 60 40 20 18 16 14 12 10 9 8 7 6 5 4

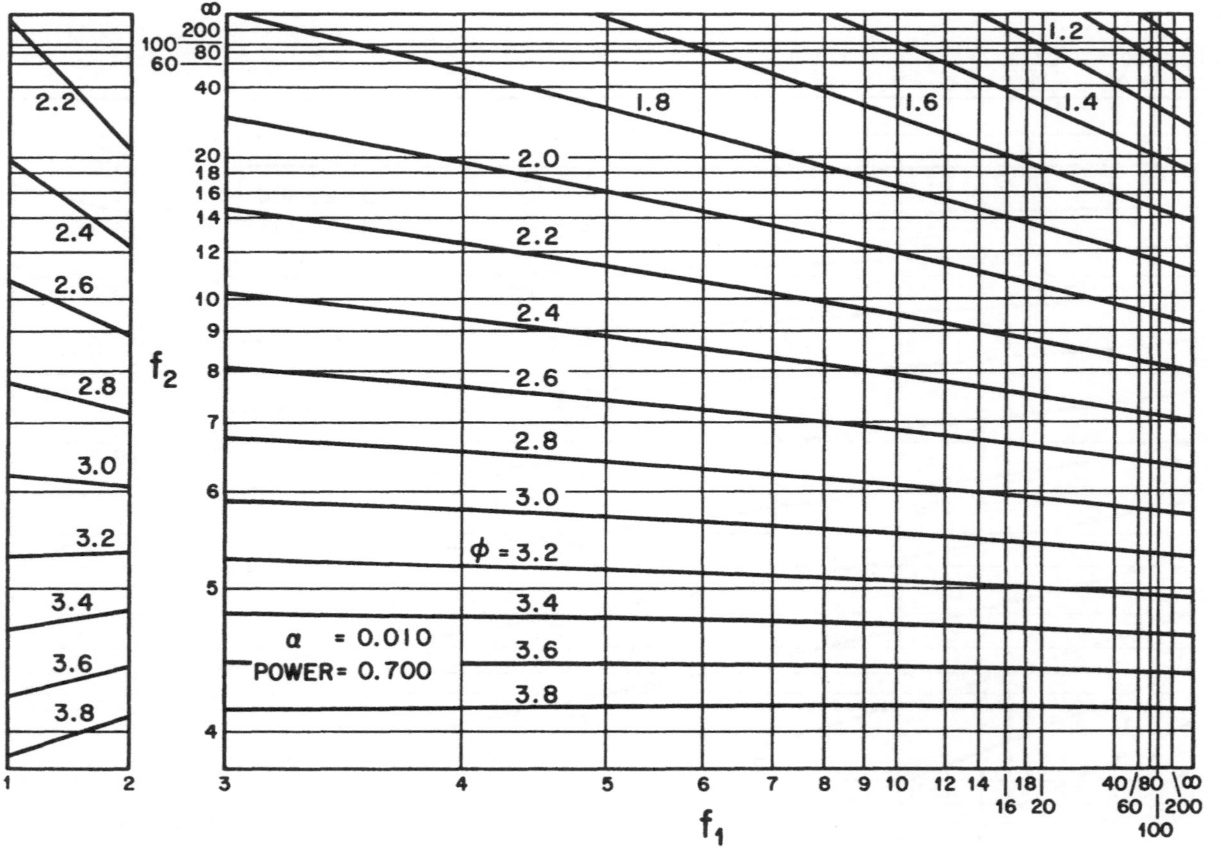
α = 0.010
POWER = 0.700
φ = 3.2
1.2
1.4
1.6
1.8
2.0
2.2
2.4
2.6
2.8
3.0
3.4
3.6
3.8
f1
f2

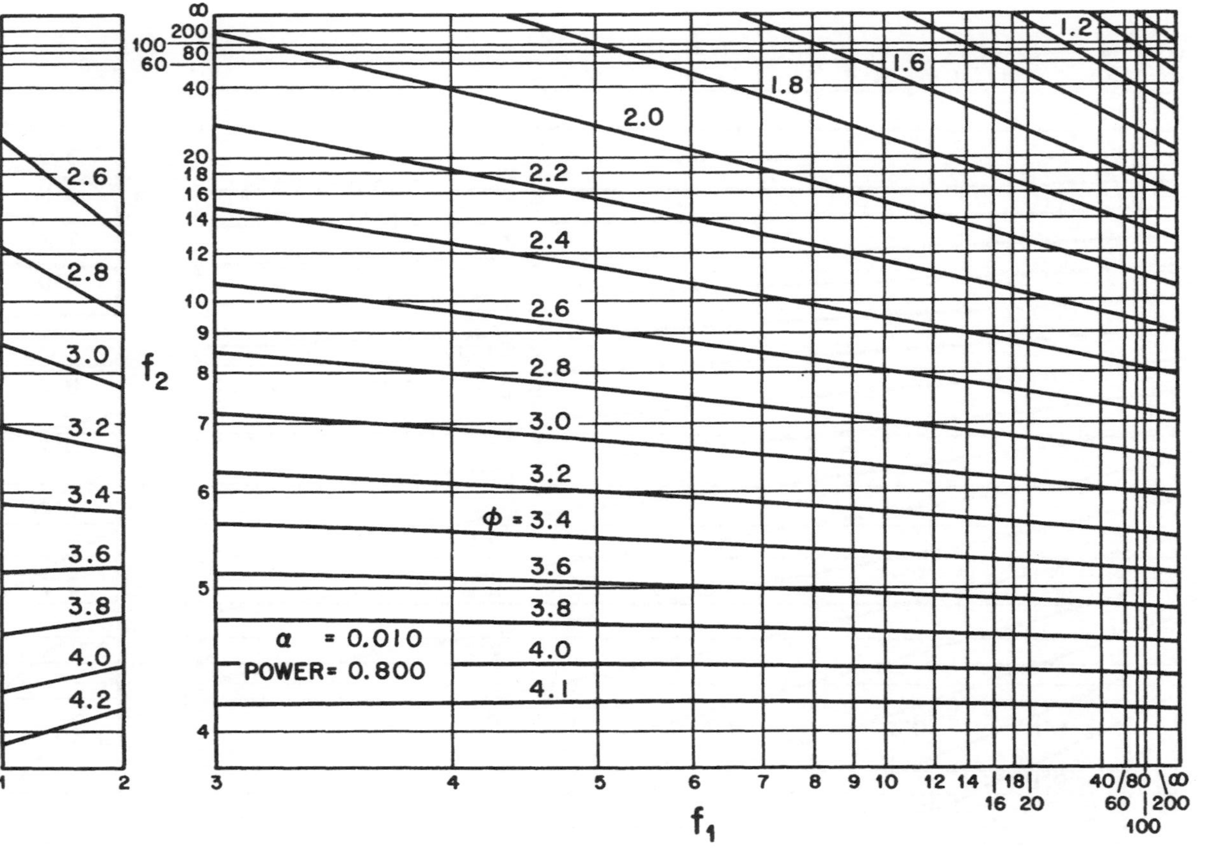

α = 0.010
POWER = 0.800
φ = 3.4
f_1
f_2

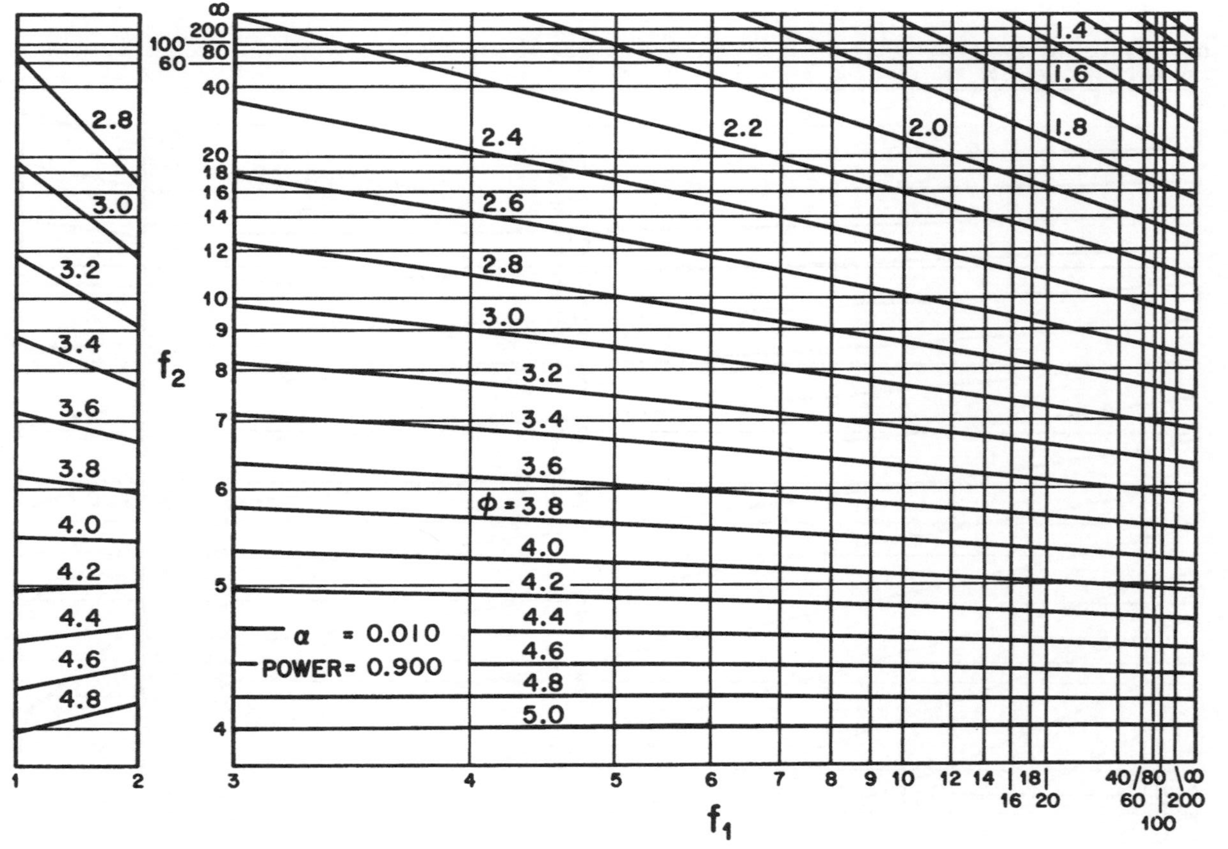
α = 0.010
POWER = 0.900
φ = 3.8
f_1
f_2

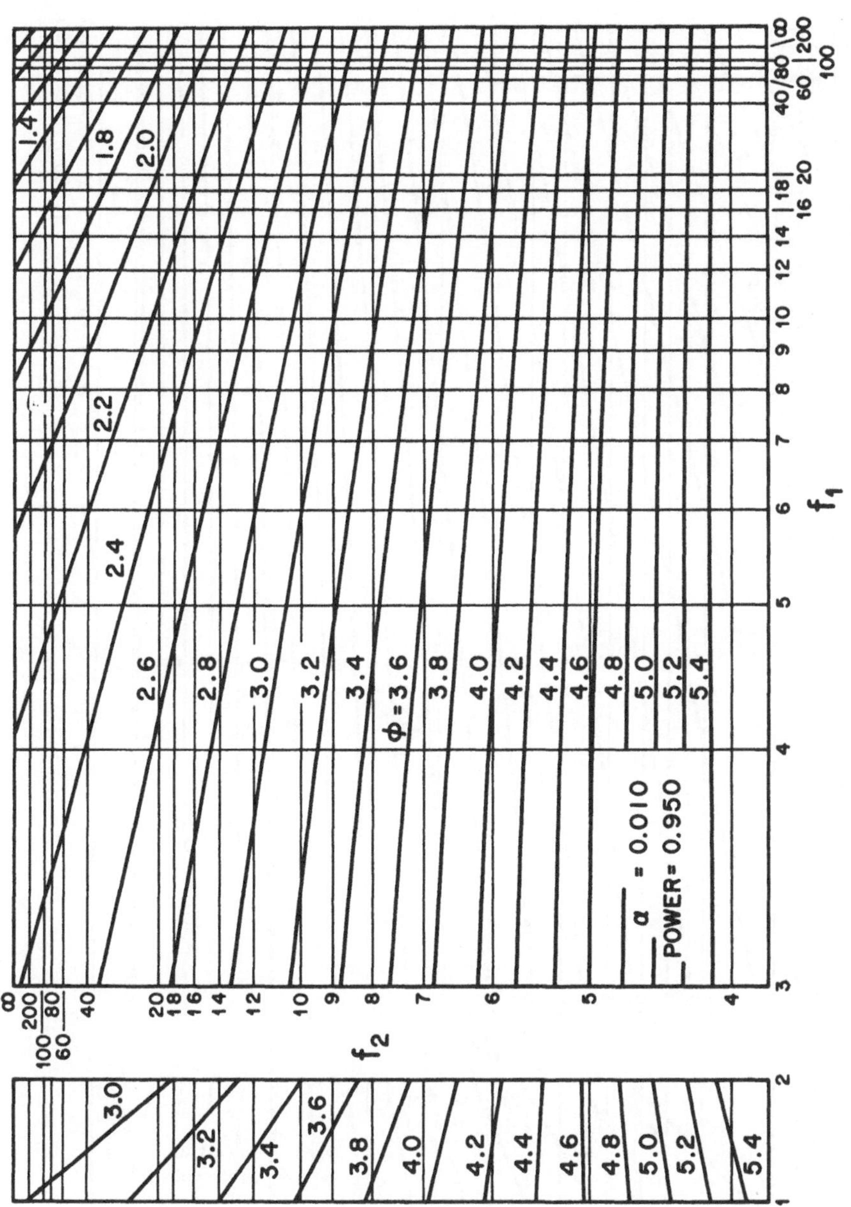

α = 0.010
POWER = 0.950
f_1
f_2
φ = 3.6
1.4
1.8
2.0
2.2
2.4
2.6
2.8
3.0
3.2
3.4
3.8
4.0
4.2
4.4
4.6
4.8
5.0
5.2
5.4

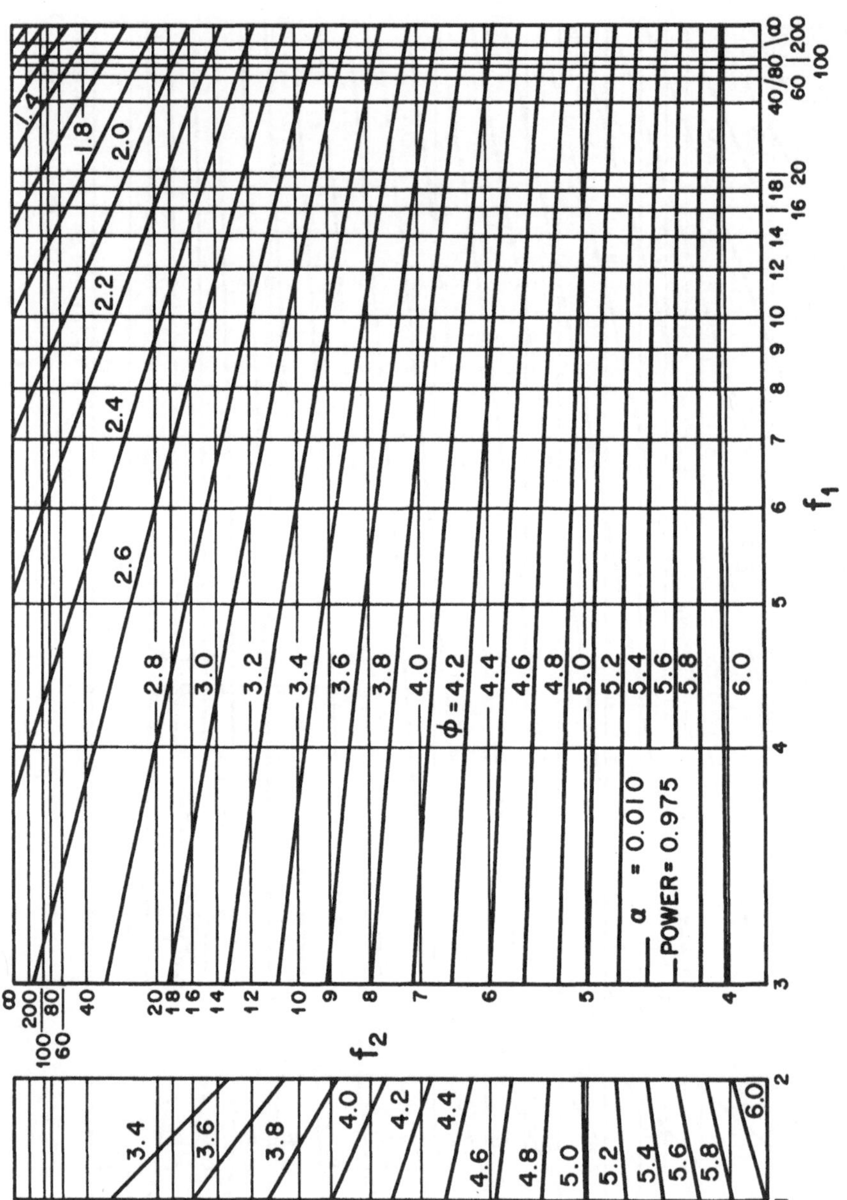
α = 0.010
POWER = 0.975
ϕ = 4.2
f_1
f_2

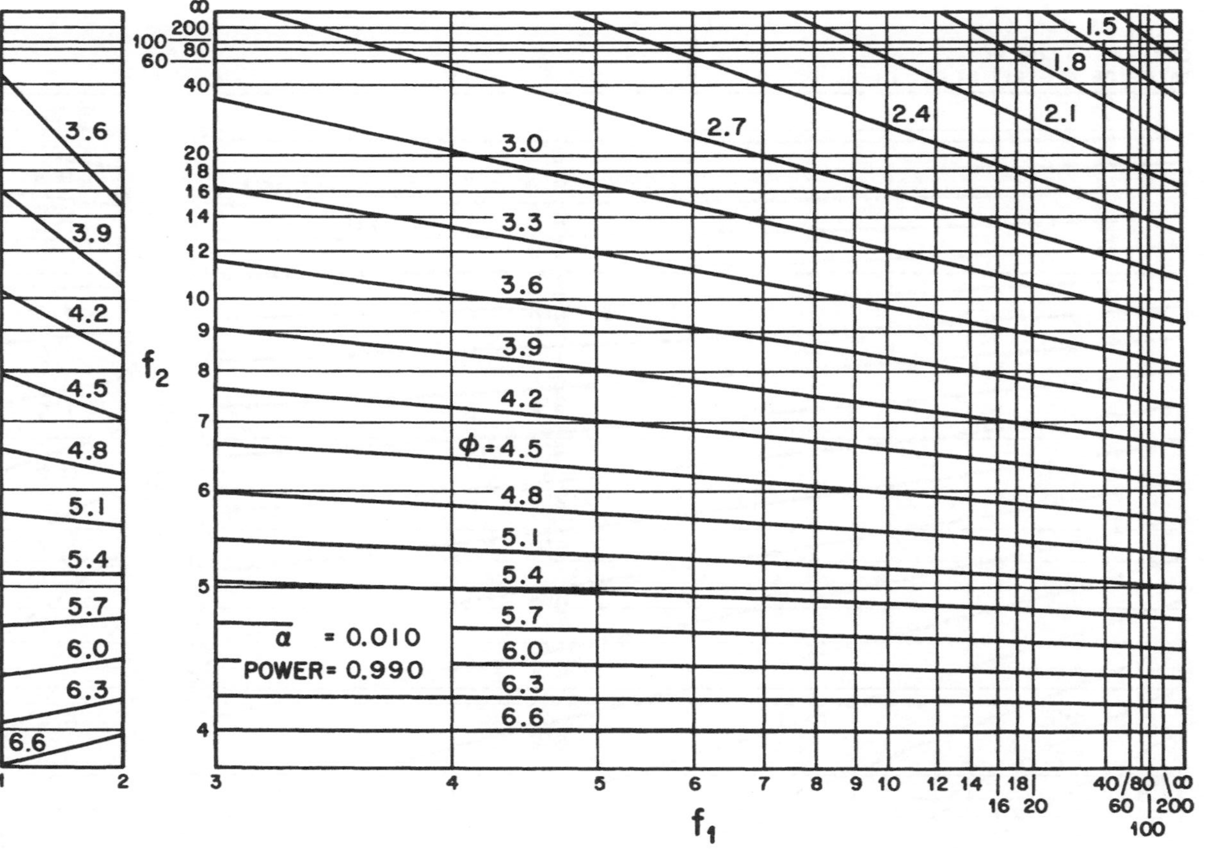

α = 0.010
POWER = 0.990
φ = 4.5
f_1
f_2

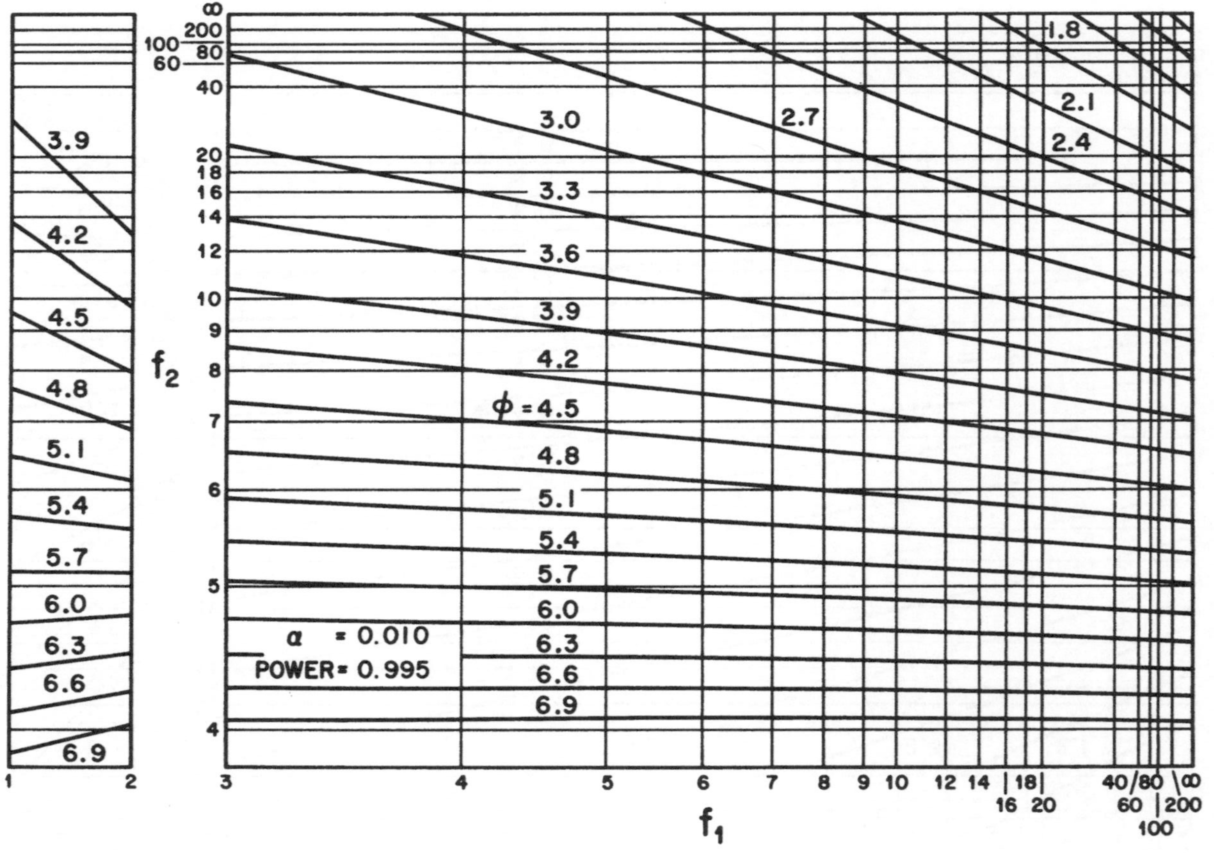

α = 0.010
POWER = 0.995
φ = 4.5
1.8
2.1
2.4
2.7
3.0
3.3
3.6
3.9
4.2
4.8
5.1
5.4
5.7
6.0
6.3
6.6
6.9
f1
f2

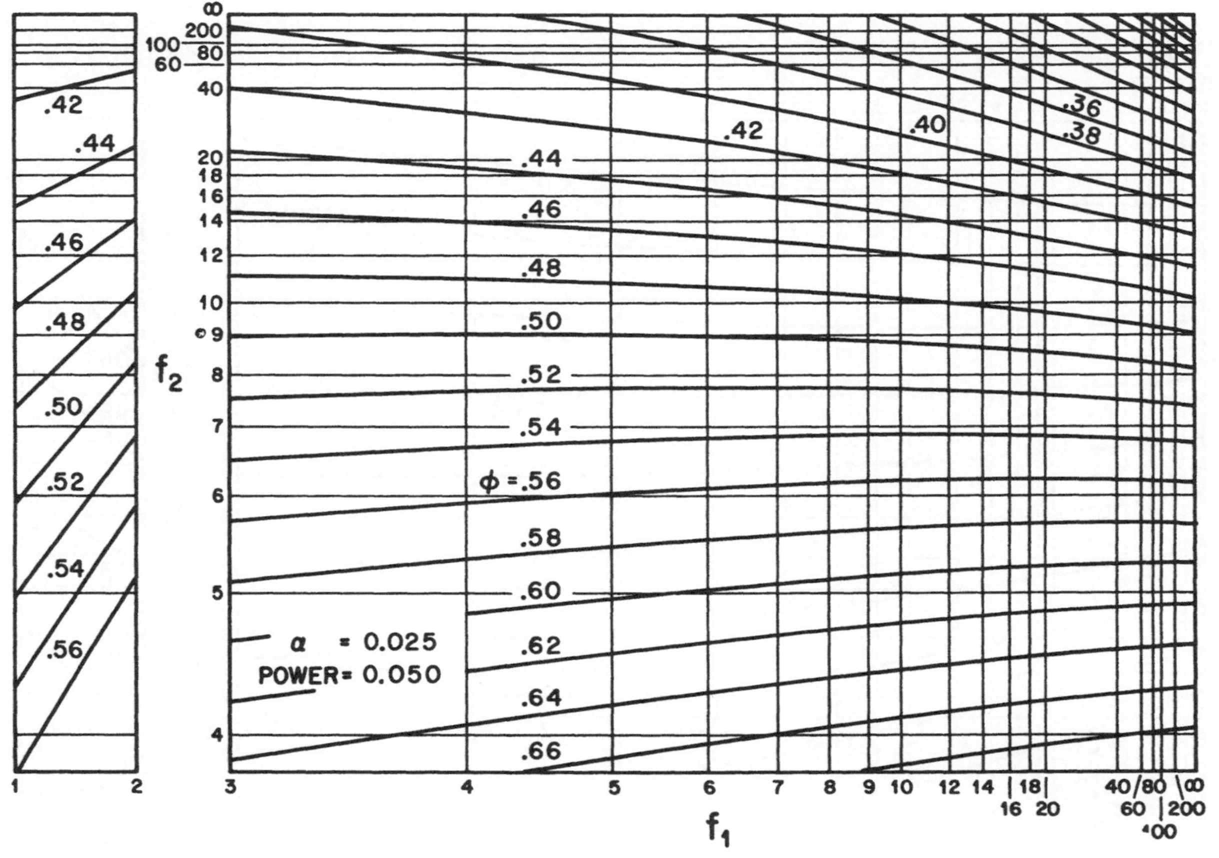
α = 0.025
POWER = 0.050
φ = .56
.36
.38
.40
.42
.44
.46
.48
.50
.52
.54
.58
.60
.62
.64
.66
f_1
f_2

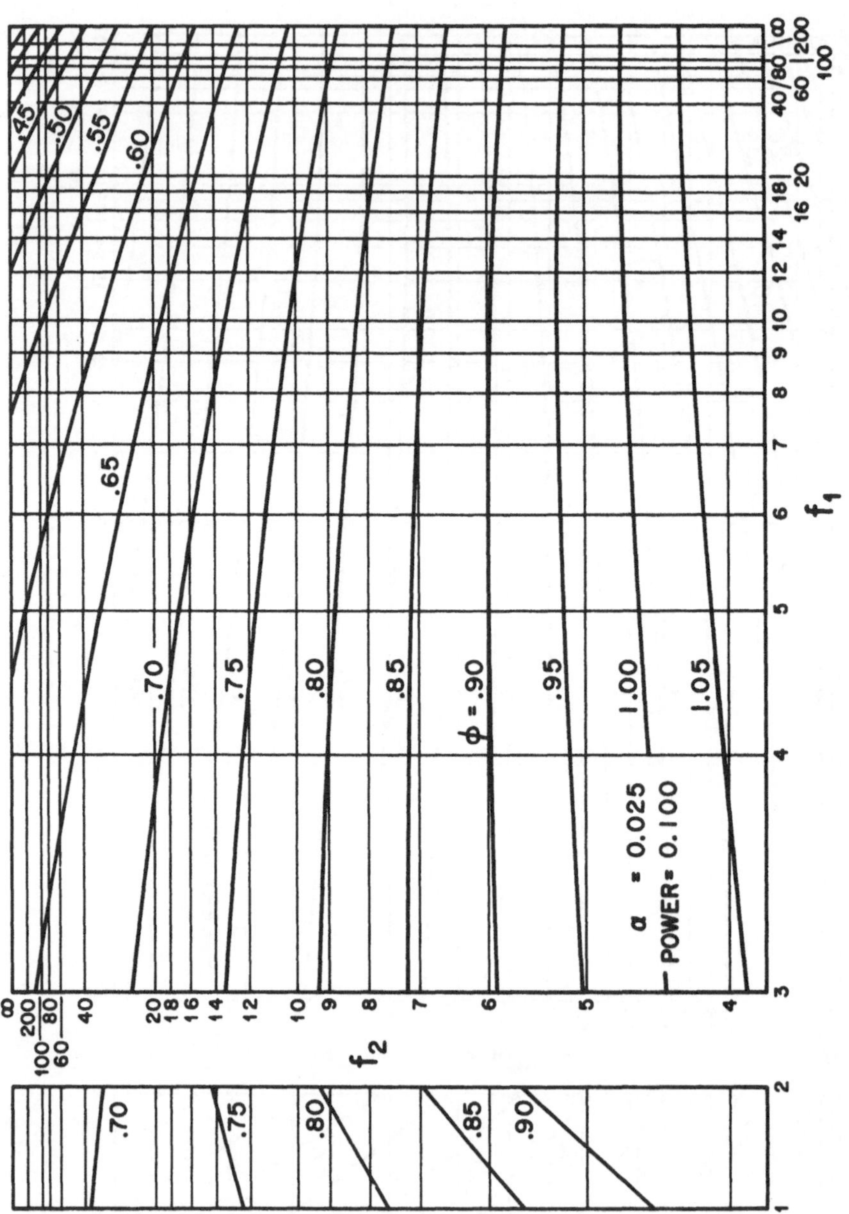

α = 0.025
POWER = 0.100
φ = .90
.45
.50
.55
.60
.65
.70
.75
.80
.85
.95
1.00
1.05
.70
.75
.80
.85
.90
f_1
f_2

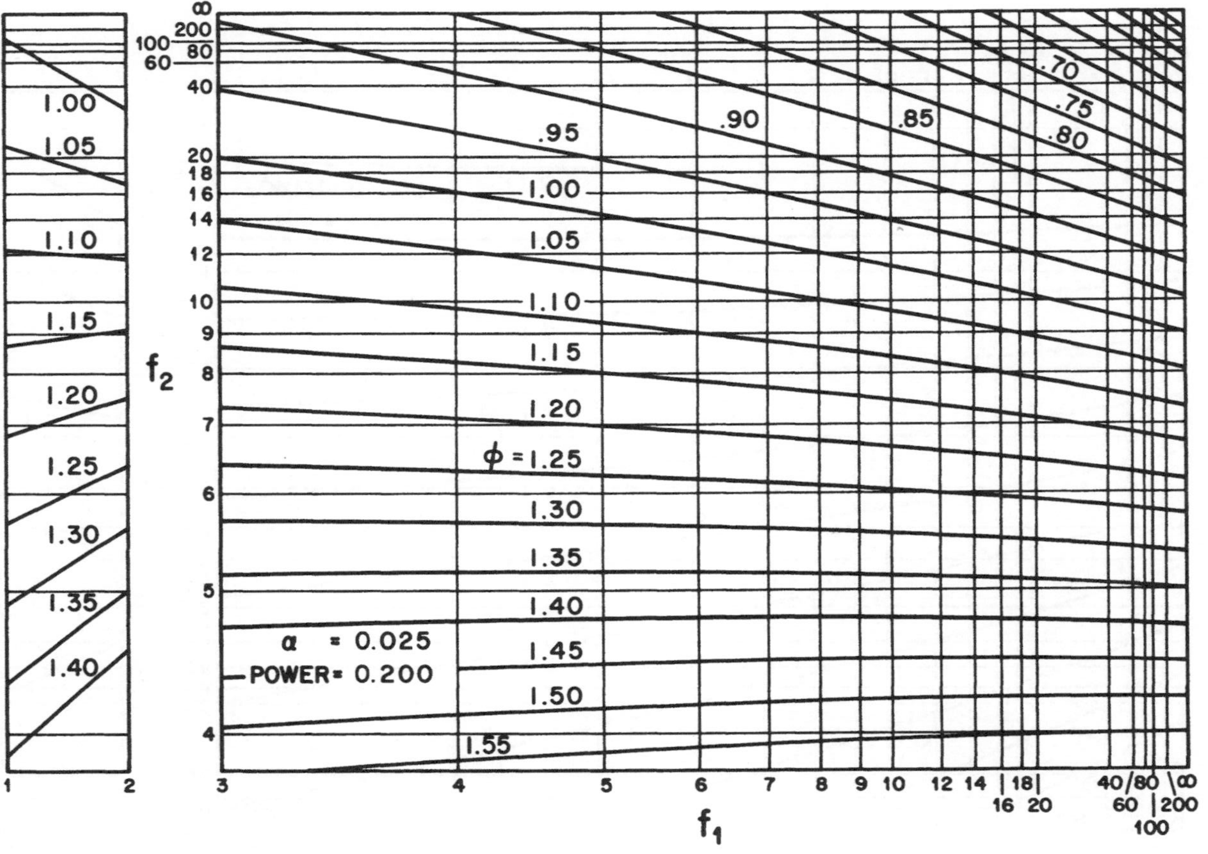

α = 0.025
POWER = 0.200
φ = 1.25
.70
.75
.80
.85
.90
.95
1.00
1.05
1.10
1.15
1.20
1.30
1.35
1.40
1.45
1.50
1.55
1.25
f_1
f_2
1 2 3 4 5 6 7 8 9 10 12 14 16 18 20 40 60 80 100 200 ∞

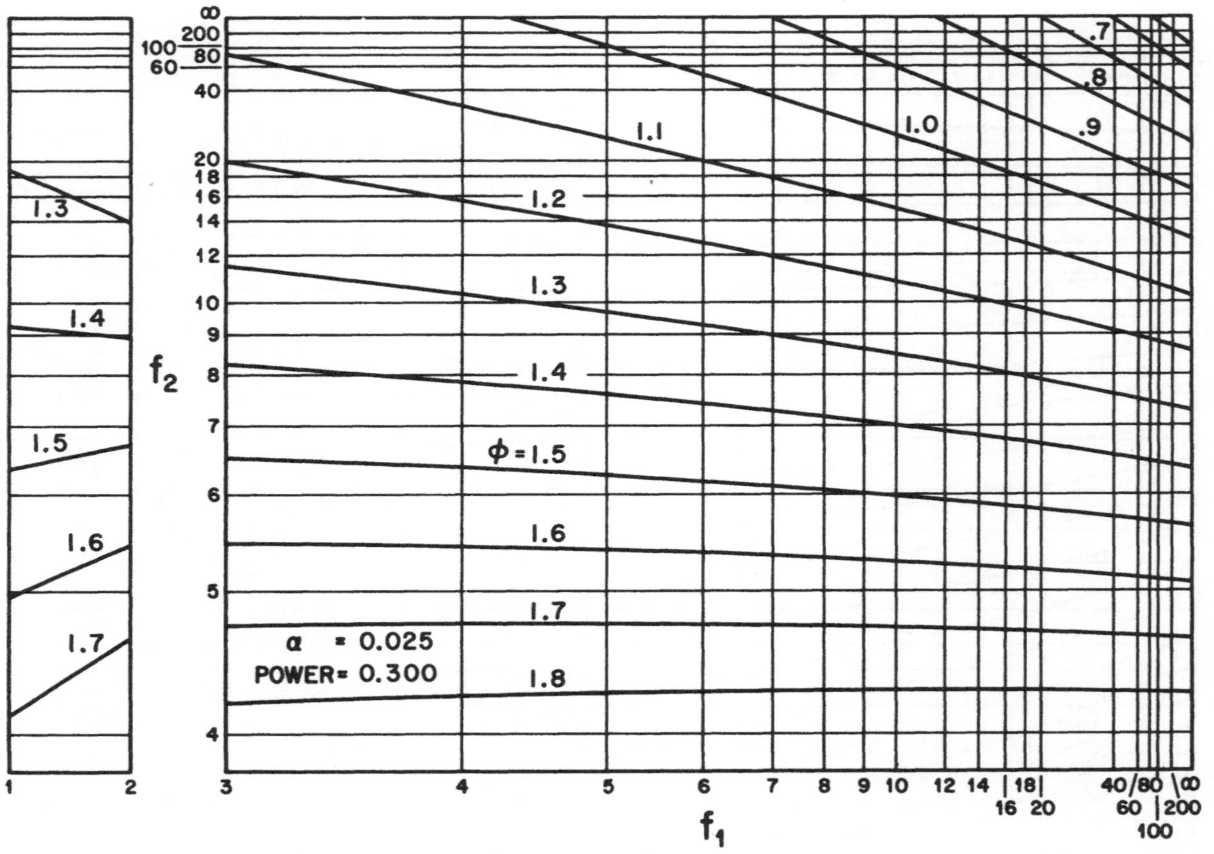

α = 0.025
POWER = 0.300
φ = 1.5
.7
.8
.9
1.0
1.1
1.2
1.3
1.4
1.6
1.7
1.8
f1
f2

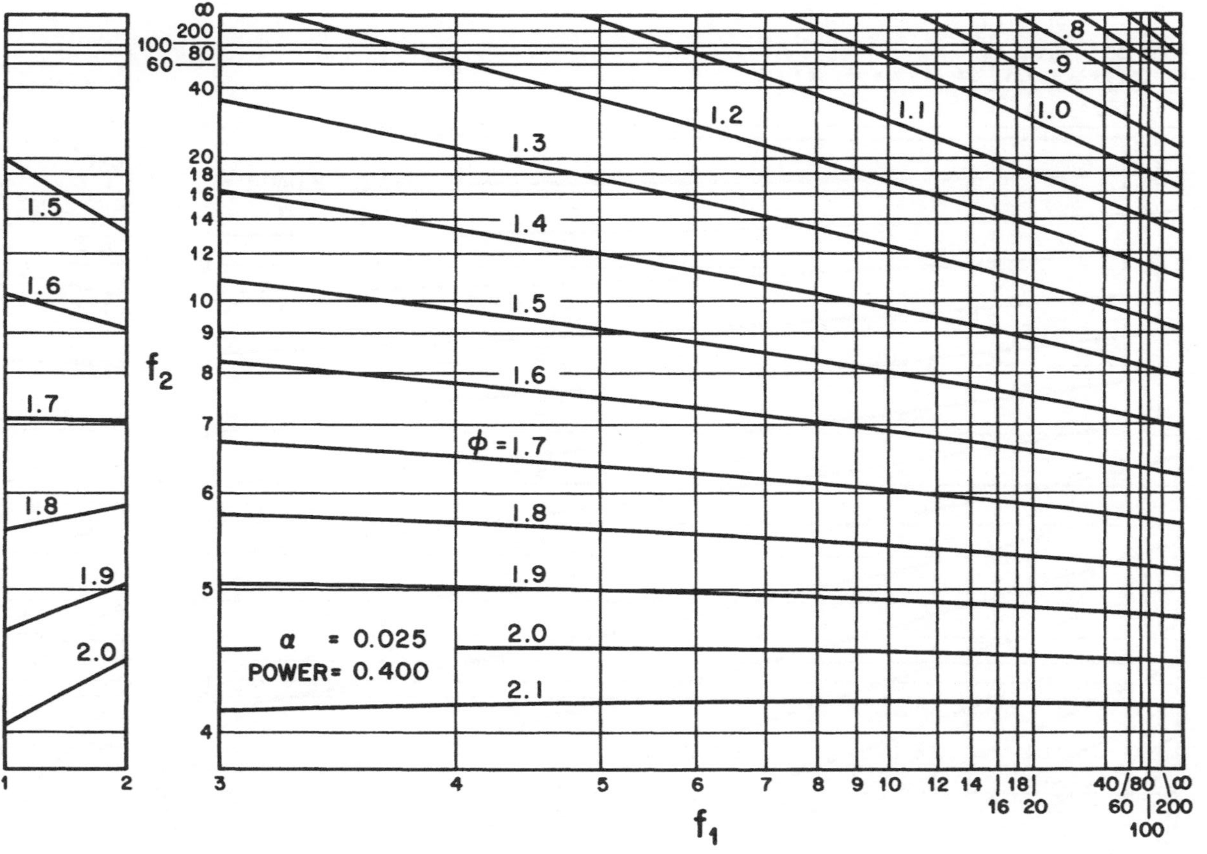
α = 0.025
POWER = 0.400
ϕ = 1.7
.8
.9
1.0
1.1
1.2
1.3
1.4
1.5
1.6
1.8
1.9
2.0
2.1
f_1
f_2

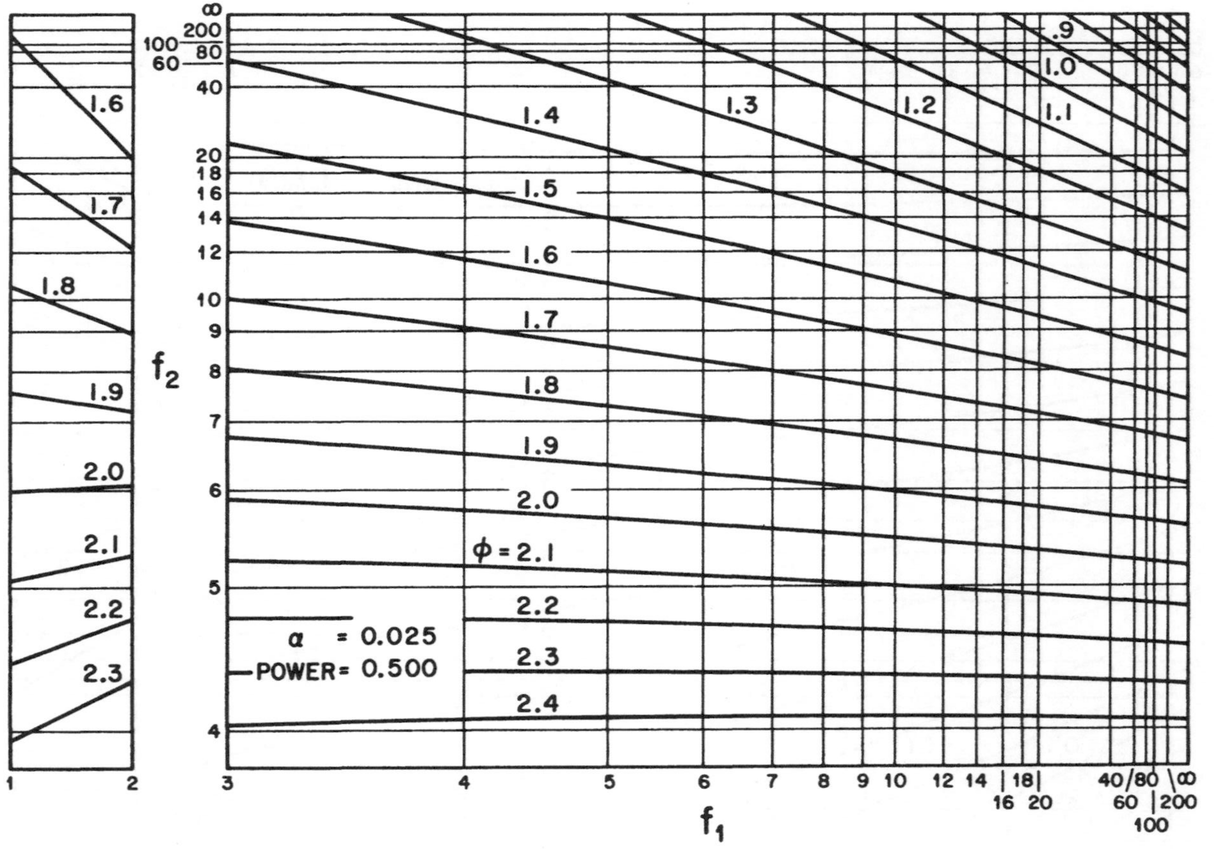

α = 0.025
POWER = 0.500
ϕ = 2.1
f_1
f_2

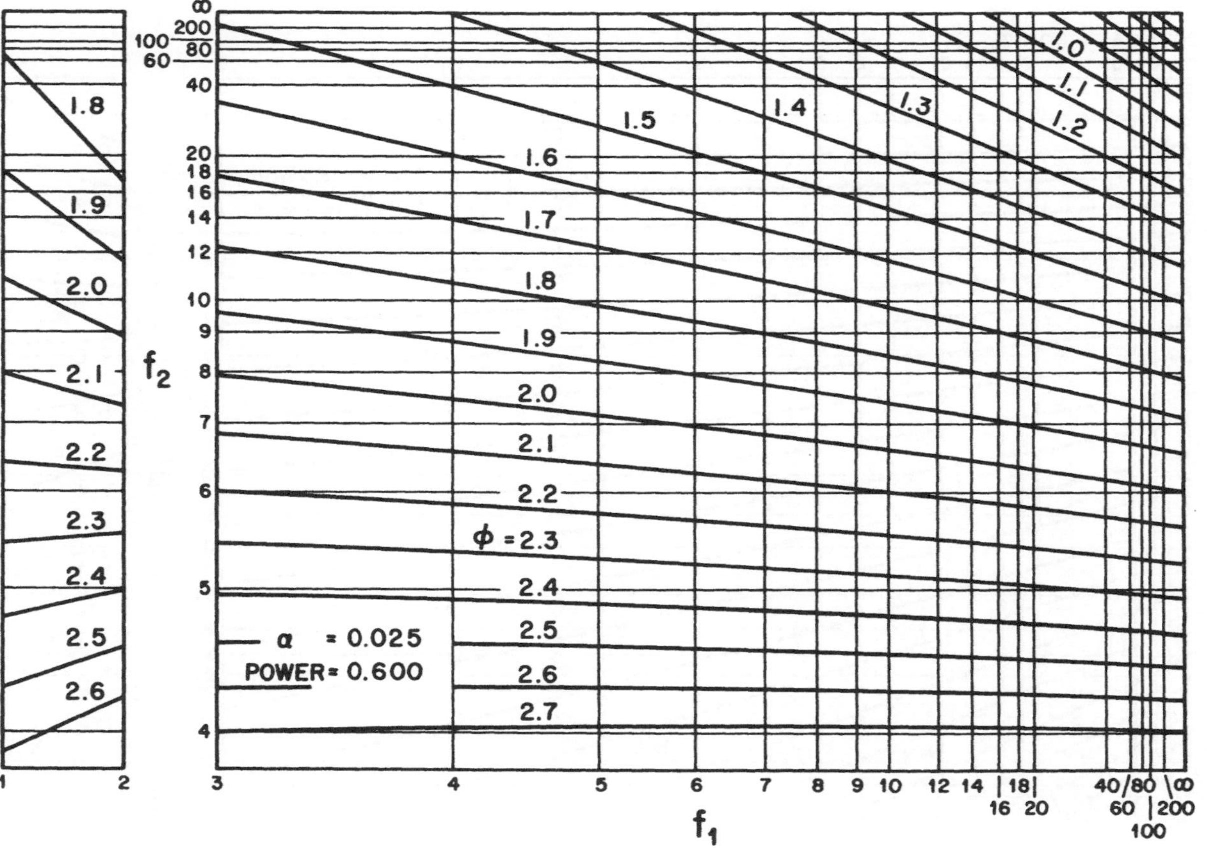

α = 0.025
POWER = 0.600
φ = 2.3
f_1
f_2

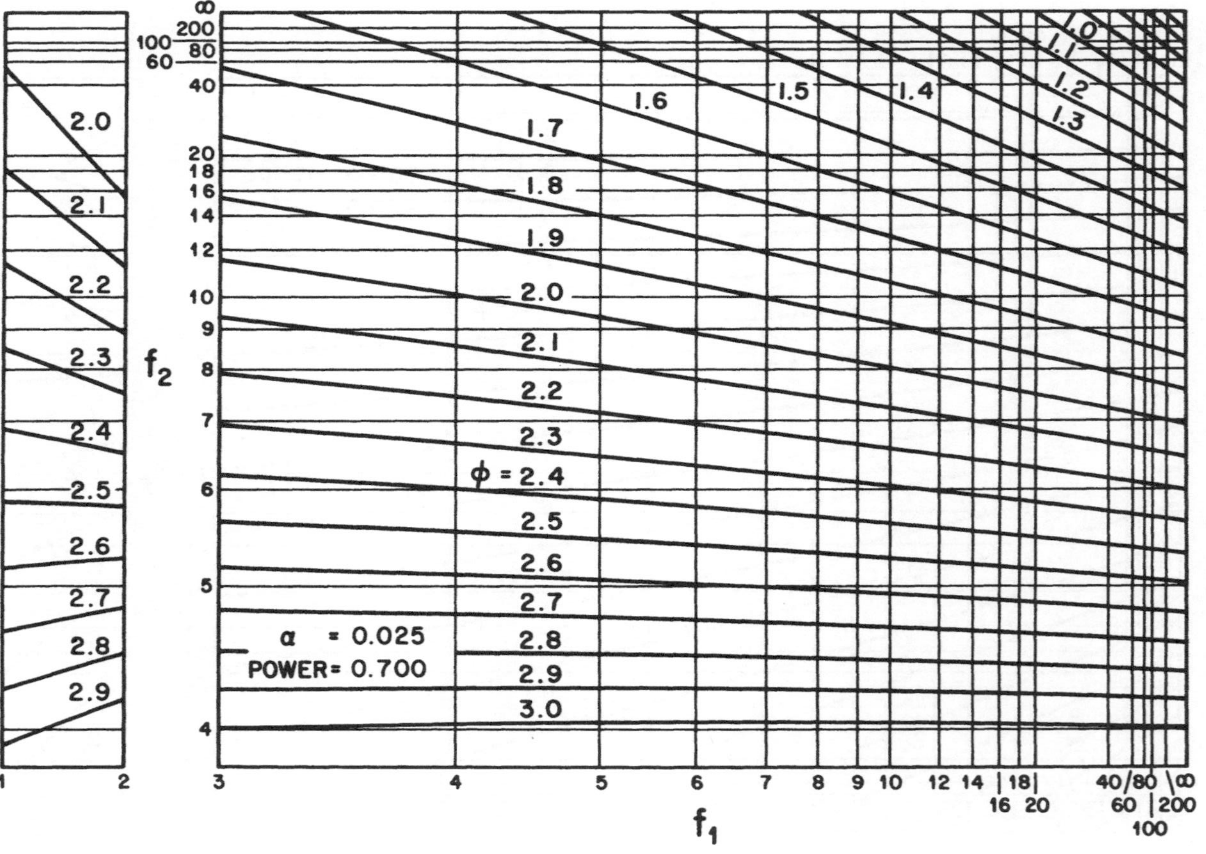
α = 0.025
POWER = 0.700
f_1
f_2
φ = 2.4
1.0
1.1
1.2
1.3
1.4
1.5
1.6
1.7
1.8
1.9
2.0
2.1
2.2
2.3
2.5
2.6
2.7
2.8
2.9
3.0

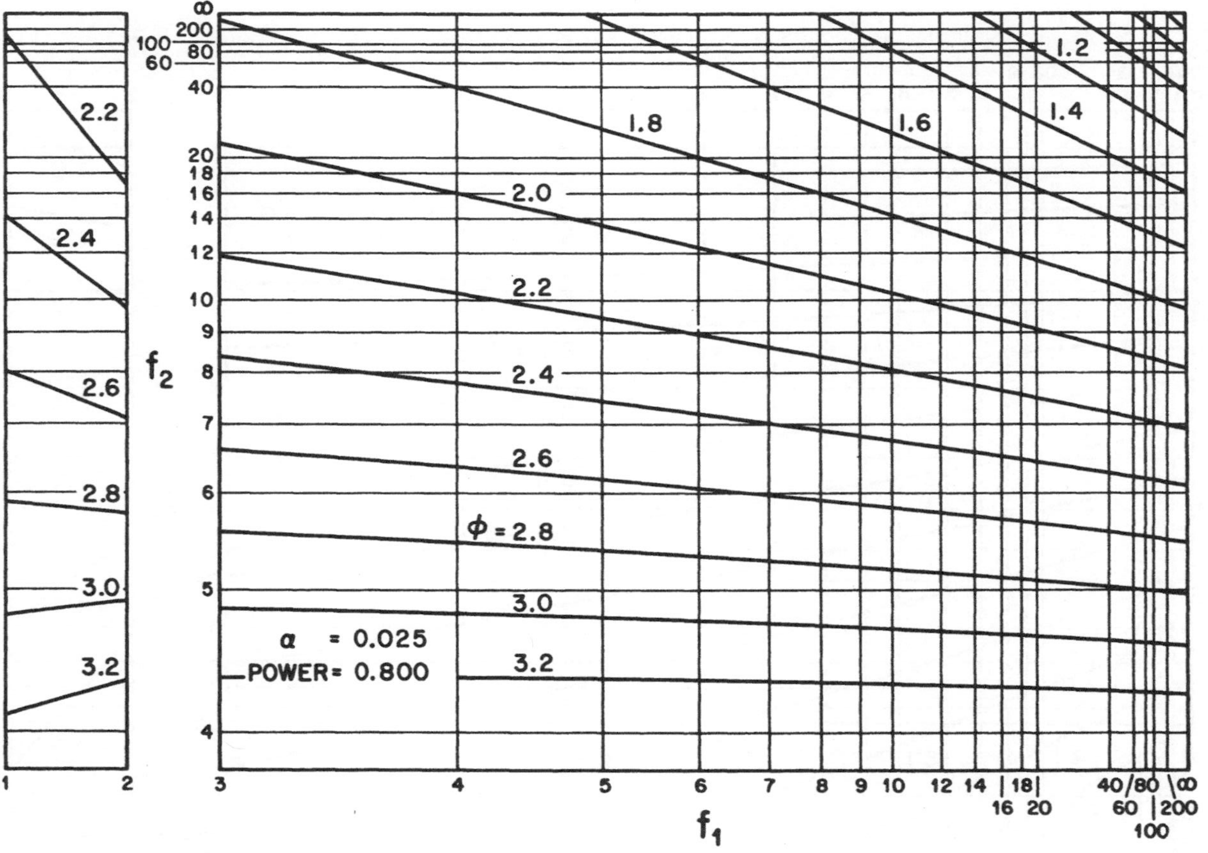

2.2
2.4
2.6
2.8
3.0
3.2
1
2
f₂
∞
200
100
80
60
40
20
18
16
14
12
10
9
8
7
6
5
4
1.2
1.4
1.6
1.8
2.0
2.2
2.4
2.6
φ = 2.8
3.0
3.2
α = 0.025
POWER = 0.800
3
4
5
6
7
8
9
10
12
14
16
18
20
40
60
80
100
200
∞
f₁

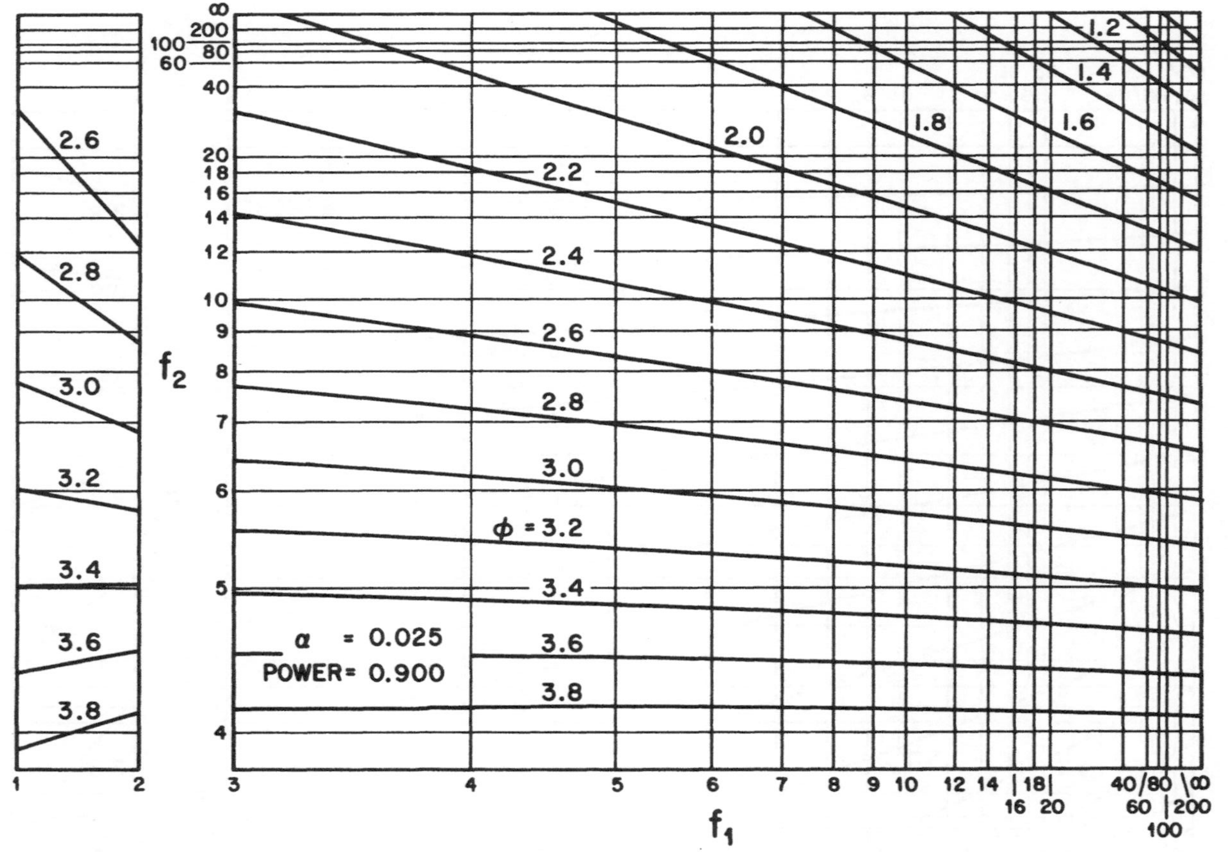
α = 0.025
POWER = 0.900
φ = 3.2
1.2
1.4
1.6
1.8
2.0
2.2
2.4
2.6
2.8
3.0
3.4
3.6
3.8
f_1
f_2

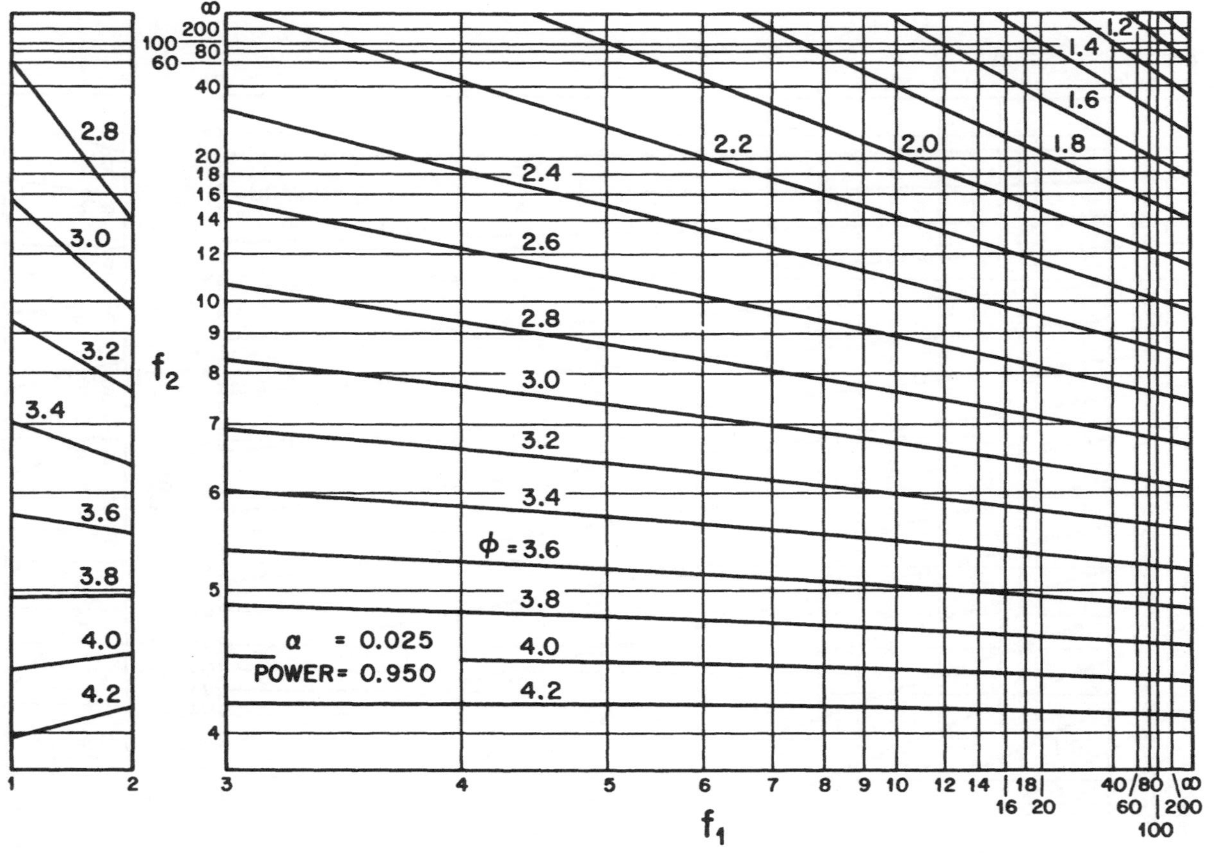

α = 0.025
POWER = 0.950
φ = 3.6
1.2
1.4
1.6
1.8
2.0
2.2
2.4
2.6
2.8
3.0
3.2
3.4
3.6
3.8
4.0
4.2
f_1
f_2

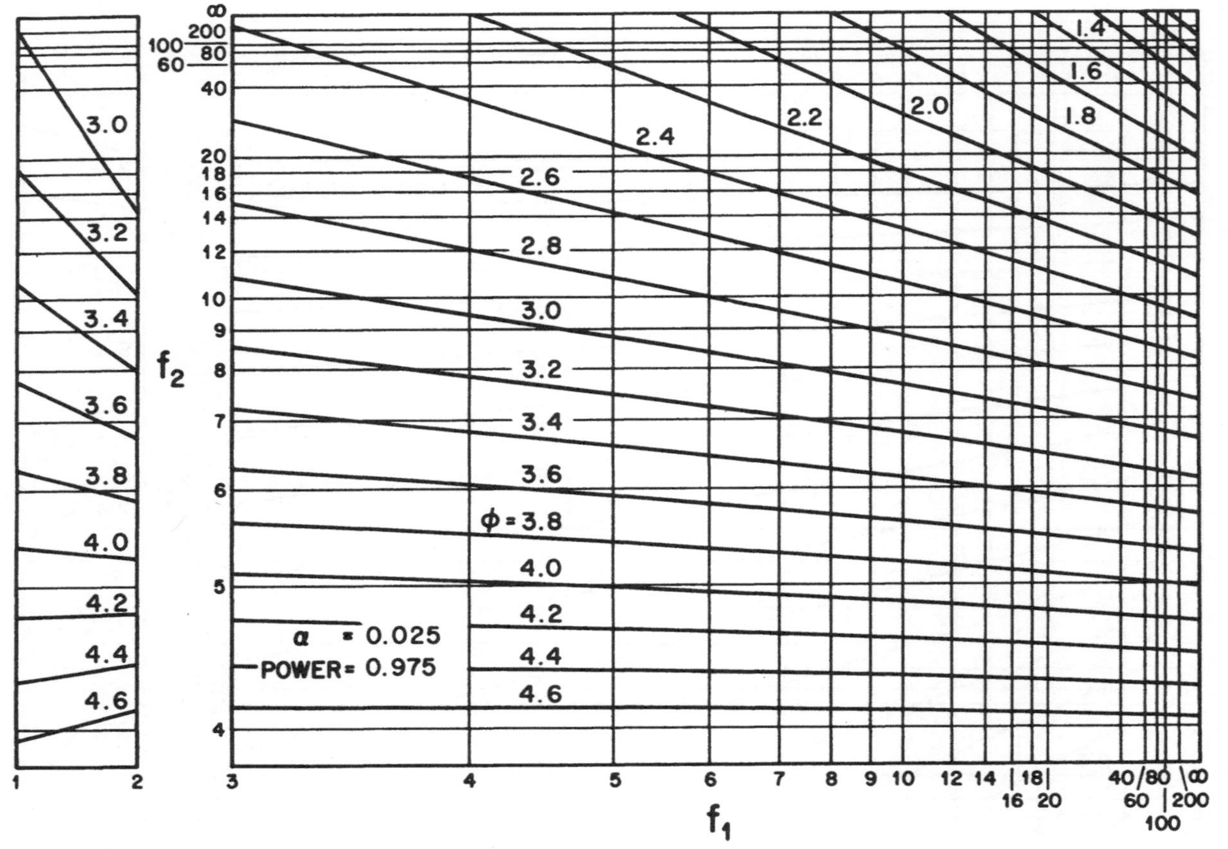

α = 0.025
POWER = 0.975
φ = 3.8
f1
f2

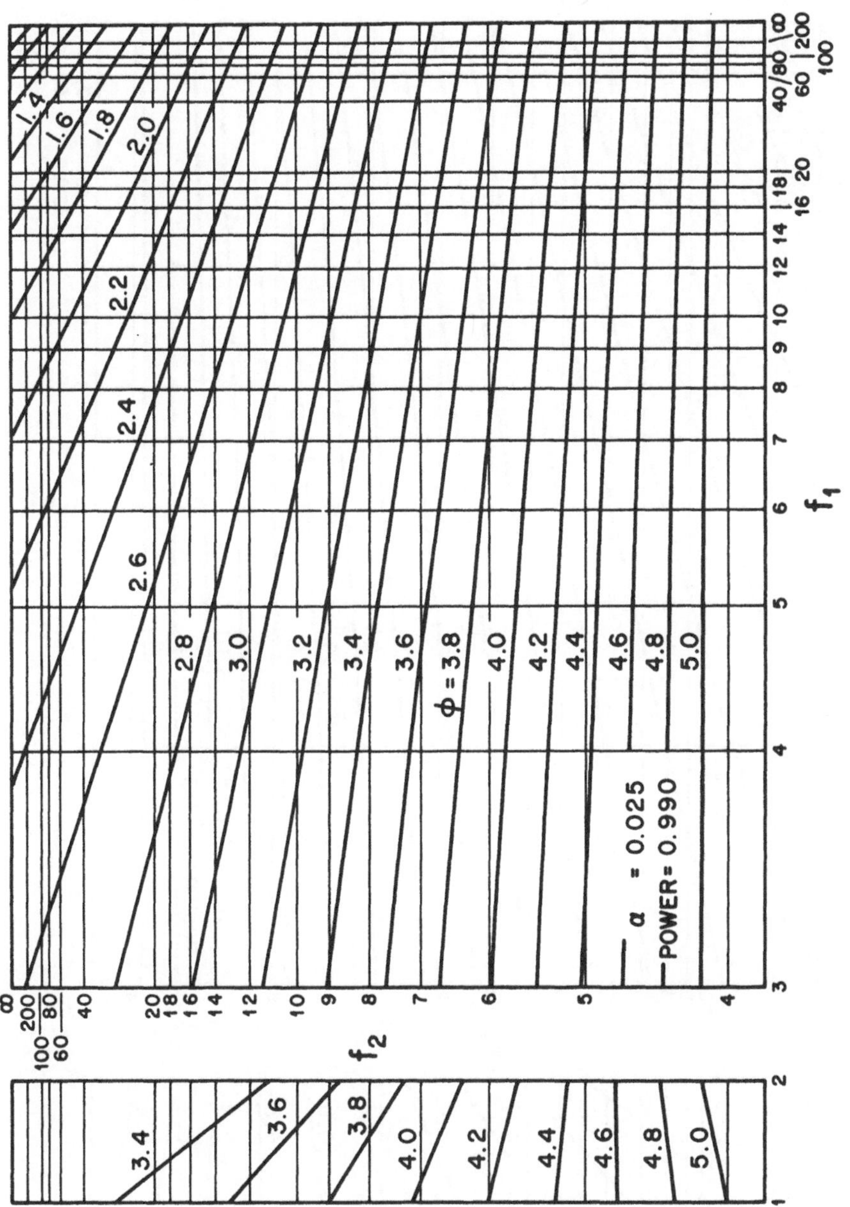

α = 0.025
POWER = 0.990
φ = 3.8
f1
f2

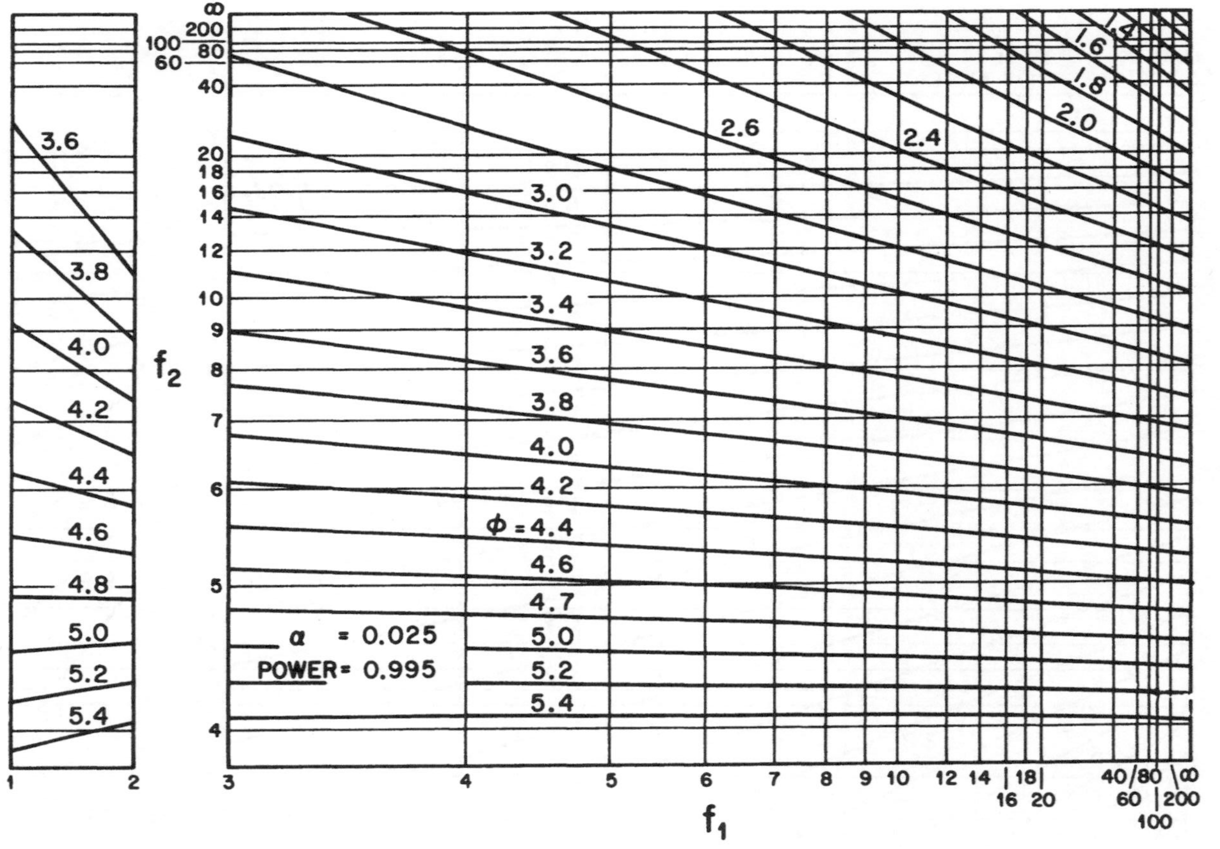

α = 0.025
POWER = 0.995
φ = 4.4
1.4
1.6
1.8
2.0
2.4
2.6
3.0
3.2
3.4
3.6
3.8
4.0
4.2
4.6
4.7
5.0
5.2
5.4
3.6
3.8
4.0
4.2
4.4
4.6
4.8
5.0
5.2
5.4
f1
f2
1
2
3
4
5
6
7
8
9
10
12
14
16
18
20
40
60
80
100
200
∞

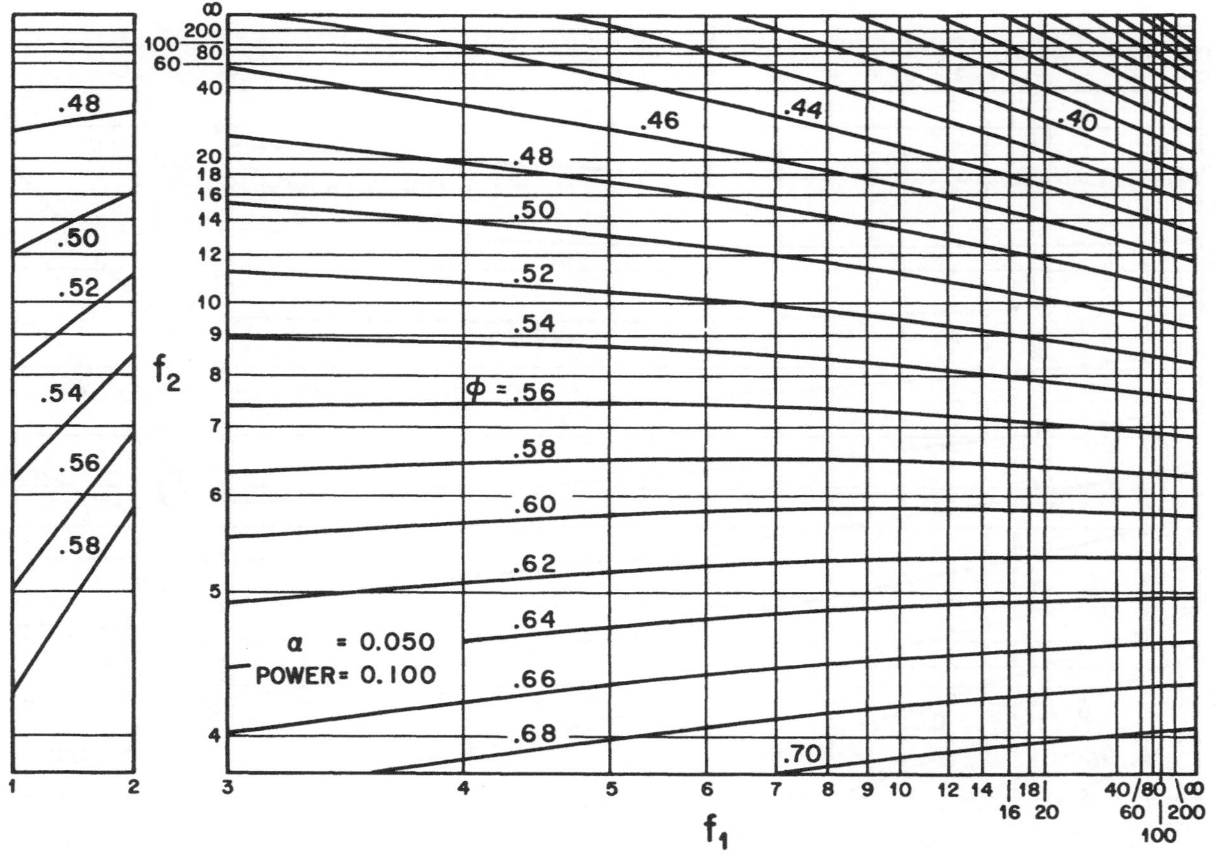

α = 0.050
POWER = 0.100
φ = .56
.40
.44
.46
.48
.50
.52
.54
.58
.60
.62
.64
.66
.68
.70
f_1
f_2

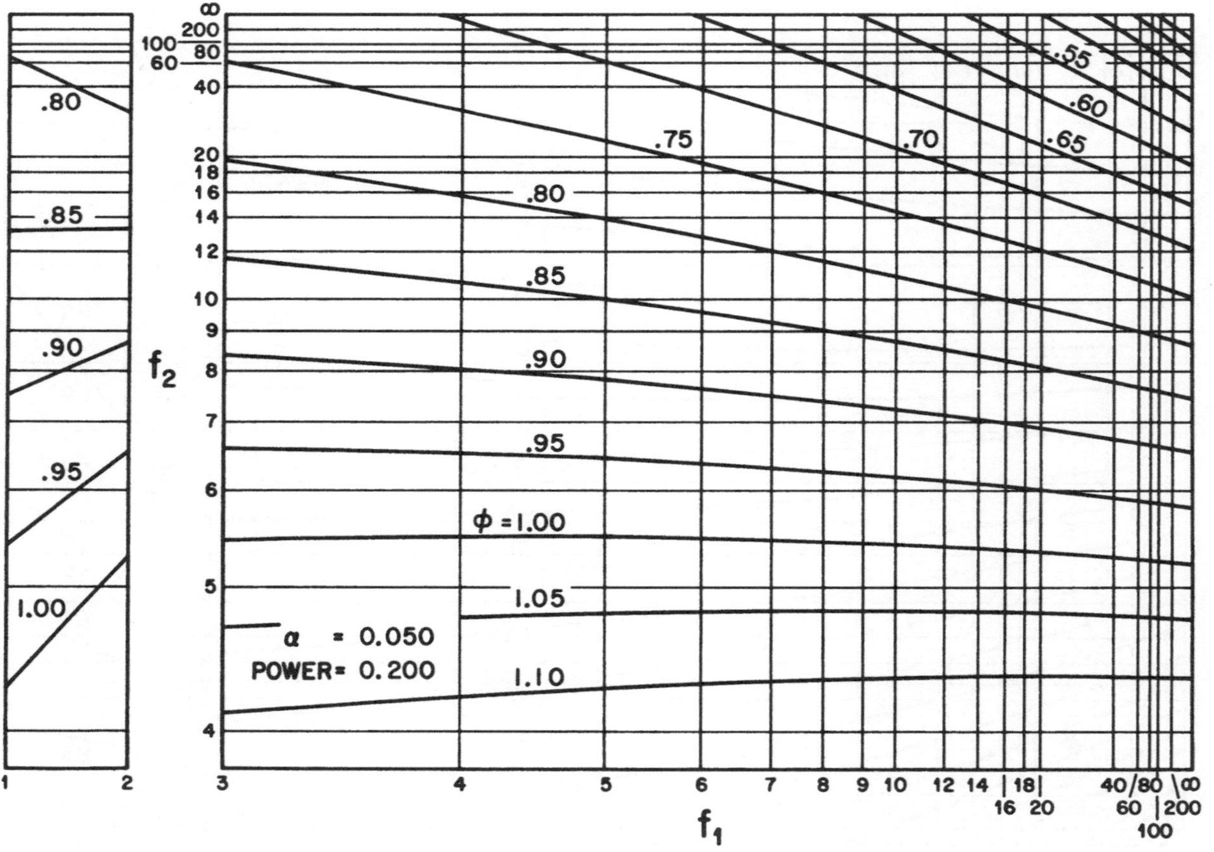

α = 0.050
POWER= 0.200
φ =1.00
.55
.60
.65
.70
.75
.80
.85
.90
.95
1.05
1.10
.80
.85
.90
.95
1.00
f_1
f_2

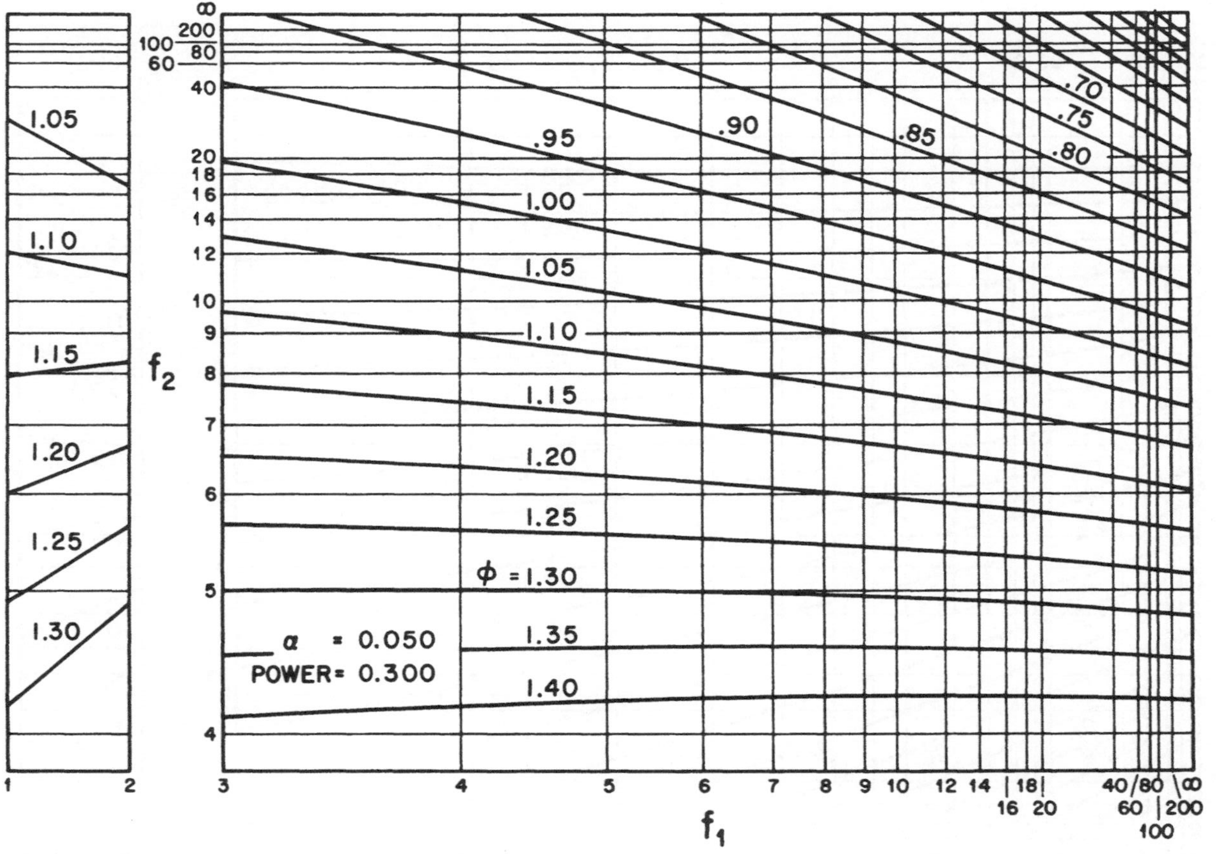
α = 0.050
POWER = 0.300
φ = 1.30
.70
.75
.80
.85
.90
.95
1.00
1.05
1.10
1.15
1.20
1.25
1.35
1.40
1.05
1.10
1.15
1.20
1.25
1.30
f_1
f_2
1
2
3
4
5
6
7
8
9
10
12
14
16
18
20
40
60
80
100
200
∞
4
5
6
7
8
9
10
12
14
16
18
20
40
60
80
100
200
∞

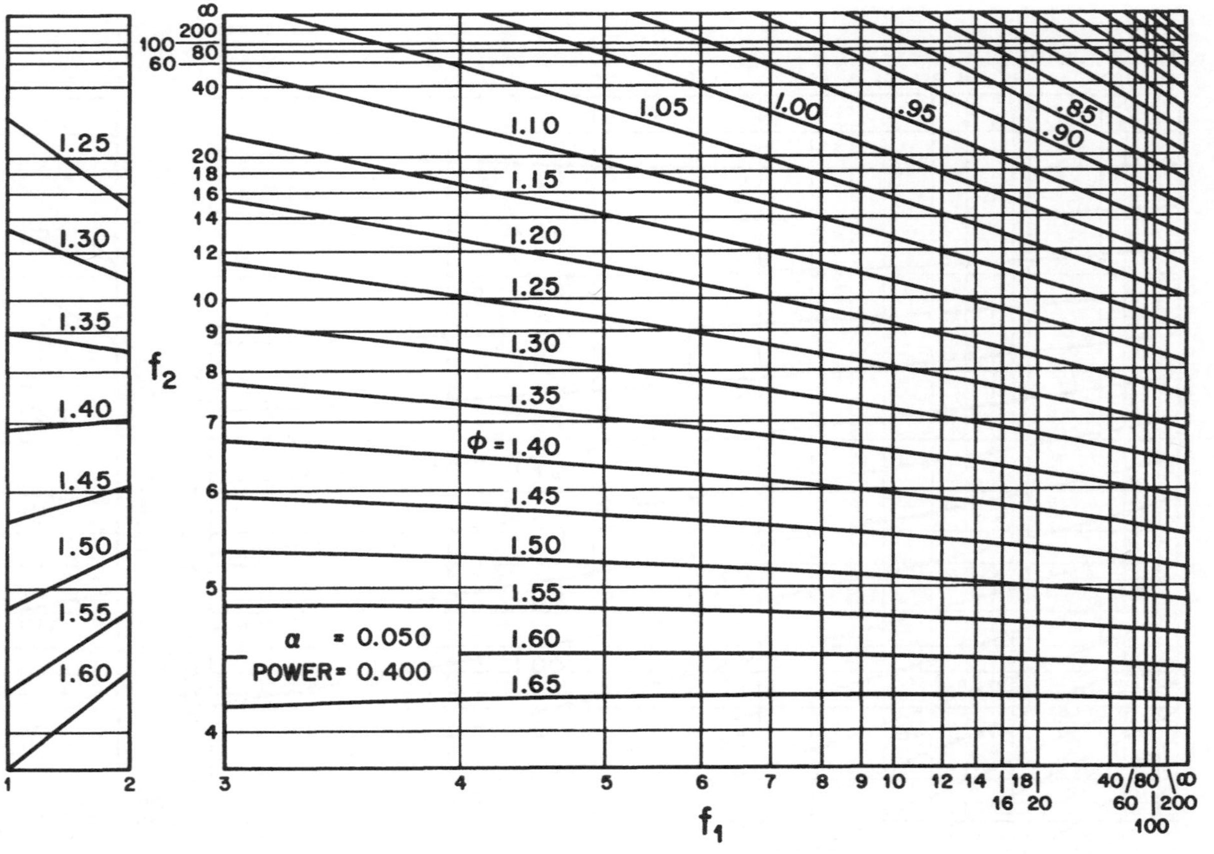

1.25
1.30
1.35
1.40
1.45
1.50
1.55
1.60
f2
1.05
1.00
.95
.85
.90
1.10
1.15
1.20
1.25
1.30
1.35
φ = 1.40
1.45
1.50
1.55
1.60
1.65
α = 0.050
POWER = 0.400
f1

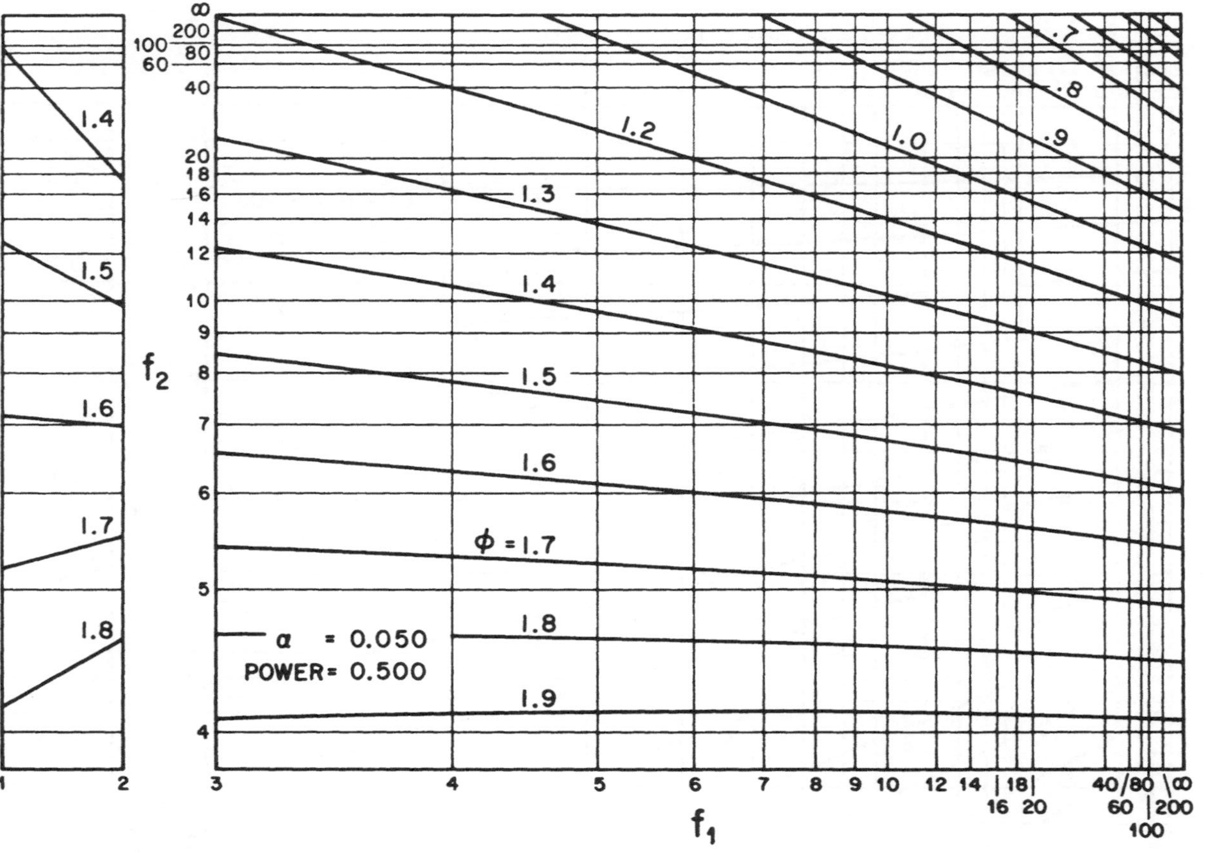

α = 0.050
POWER = 0.500
φ = 1.7
.7
.8
.9
1.0
1.2
1.3
1.4
1.5
1.6
1.8
1.9
f1
f2

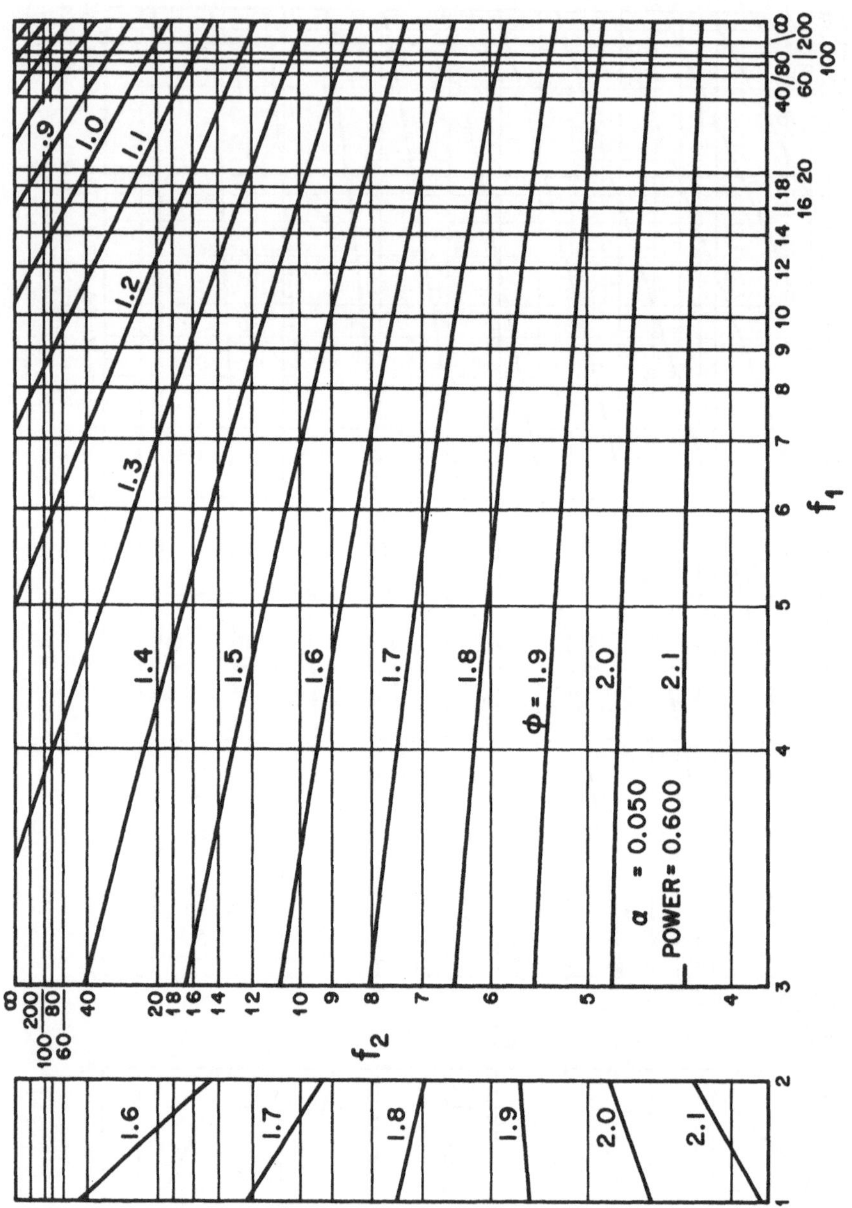

α = 0.050
POWER = 0.600
φ = 1.9
f_1
f_2

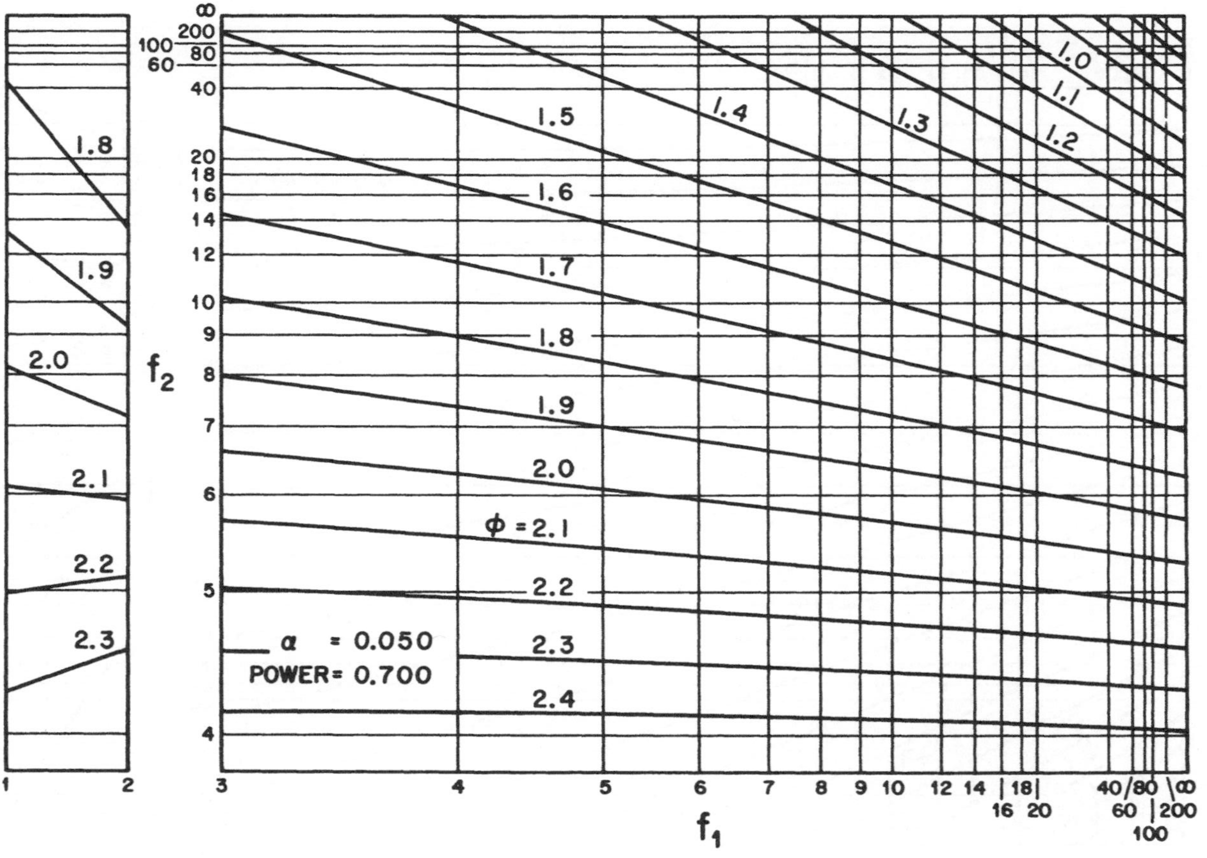
α = 0.050
POWER = 0.700
φ = 2.1
f1
f2
1.0
1.1
1.2
1.3
1.4
1.5
1.6
1.7
1.8
1.9
2.0
2.2
2.3
2.4

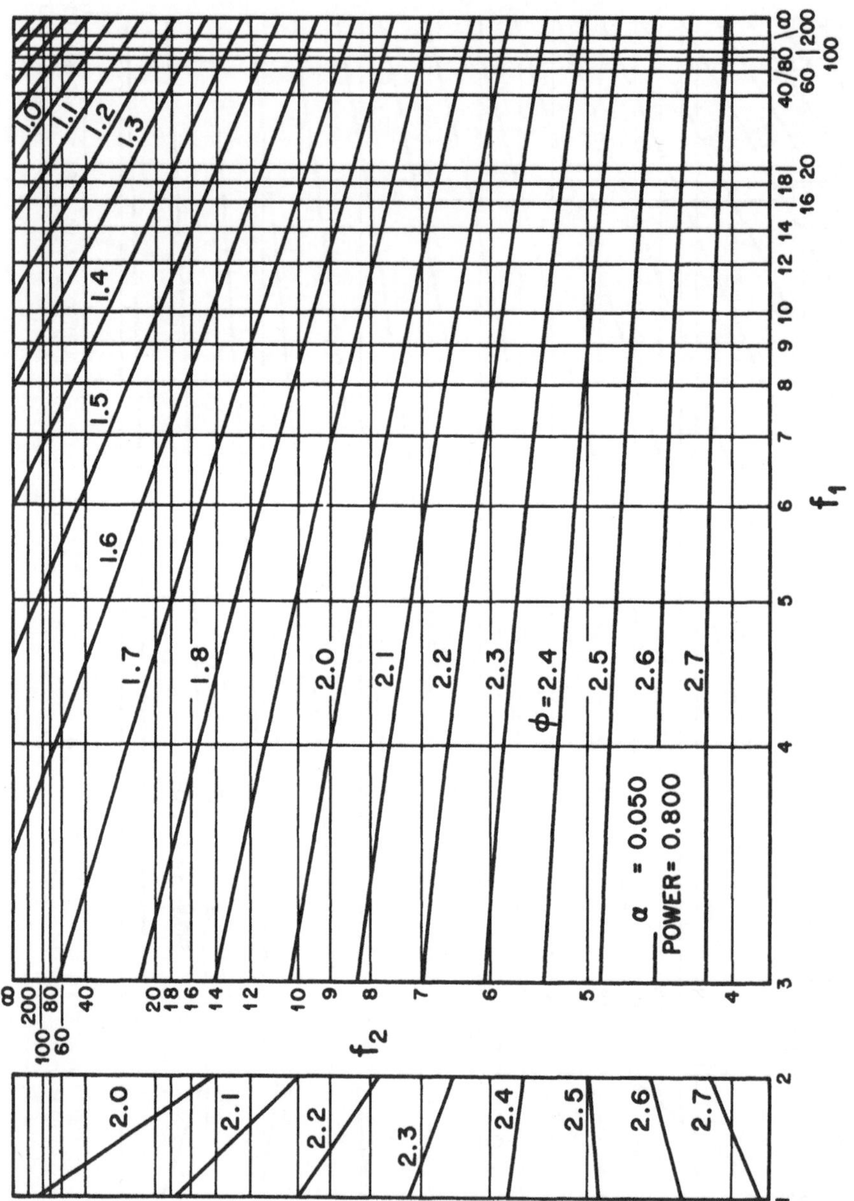
α = 0.050
POWER = 0.800
φ = 2.4
f_1
f_2

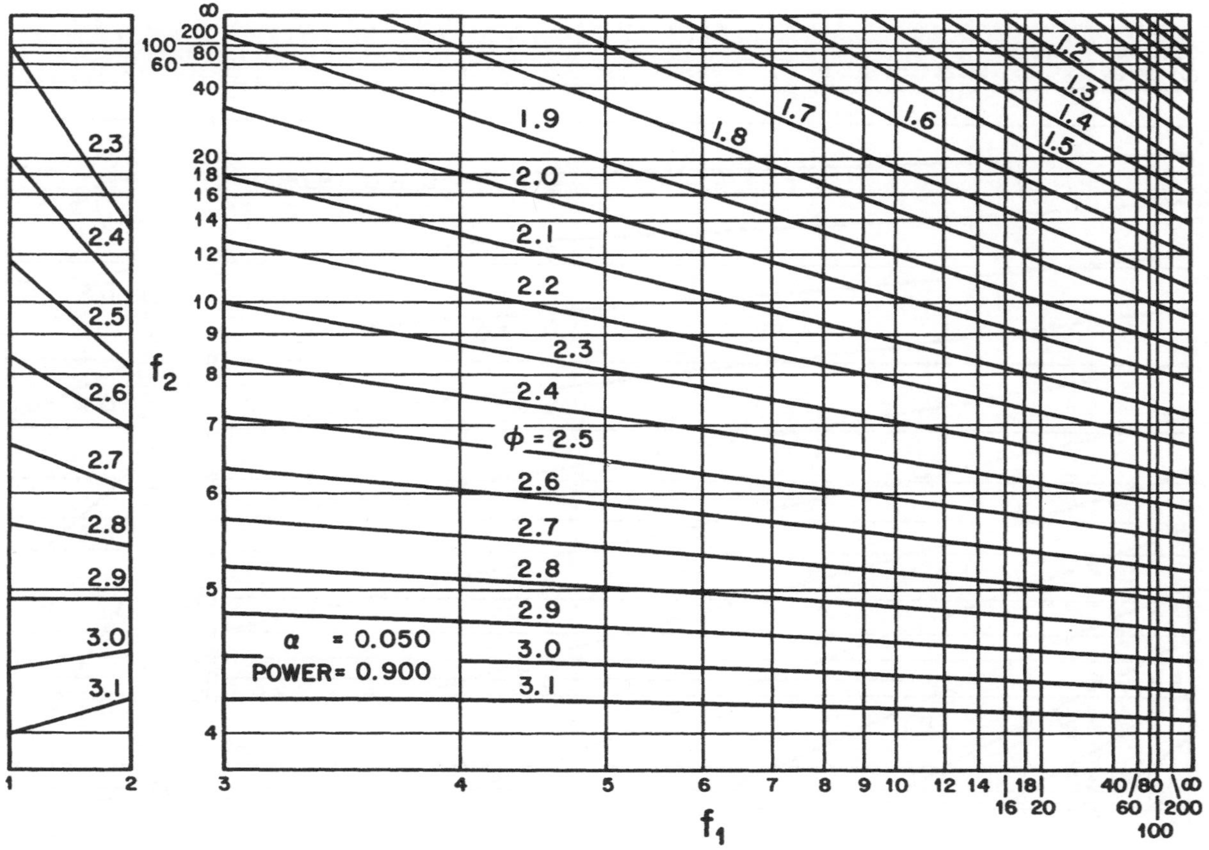

f₁
f₂
φ = 2.5
1.2
1.3
1.4
1.5
1.6
1.7
1.8
1.9
2.0
2.1
2.2
2.3
2.4
2.6
2.7
2.8
2.9
3.0
3.1
α = 0.050
POWER= 0.900
1 2 3 4 5 6 7 8 9 10 12 14 16 18 20 40 60 80 100 200 ∞
4 5 6 7 8 9 10 12 14 16 18 20 40 60 80 100 200 ∞

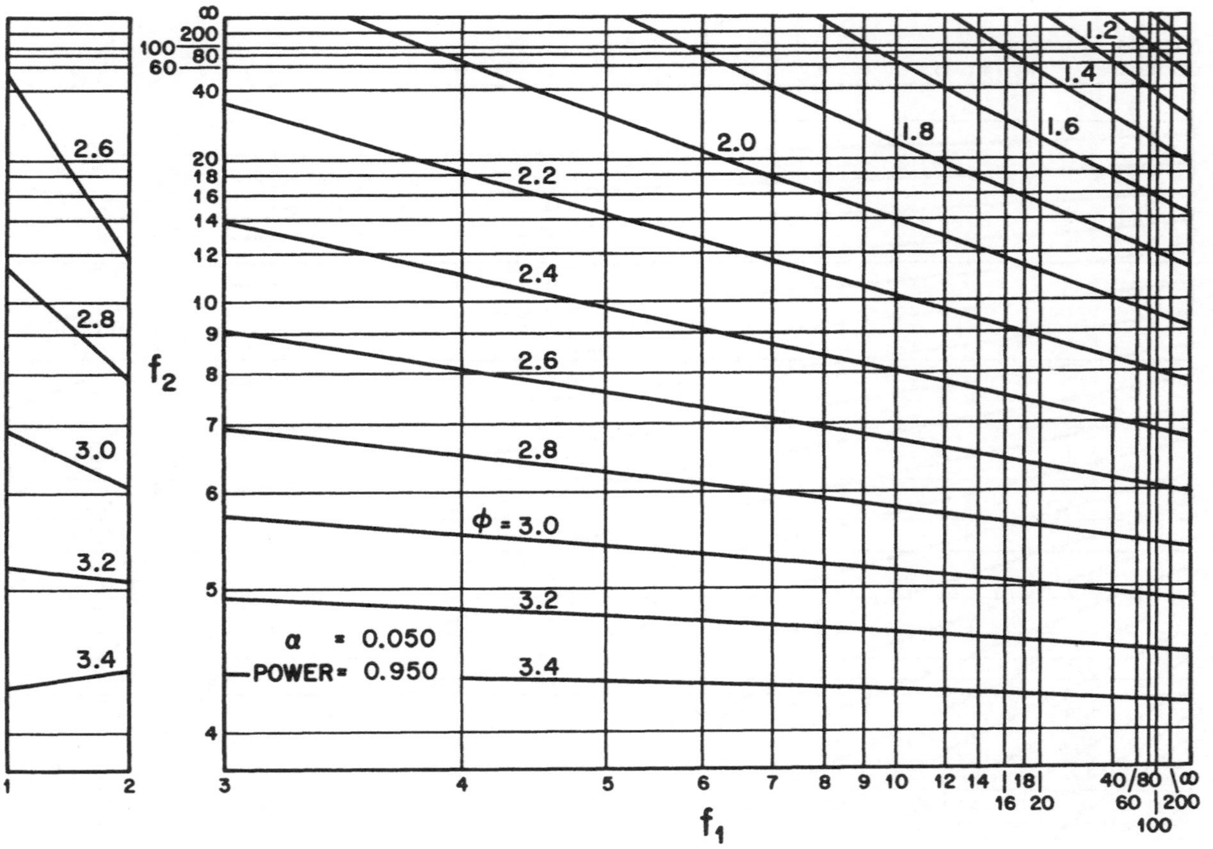

2.6
2.8
3.0
3.2
3.4
1.2
1.4
1.6
1.8
2.0
2.2
2.4
2.6
2.8
ϕ = 3.0
3.2
3.4
α = 0.050
POWER = 0.950
f_1
f_2

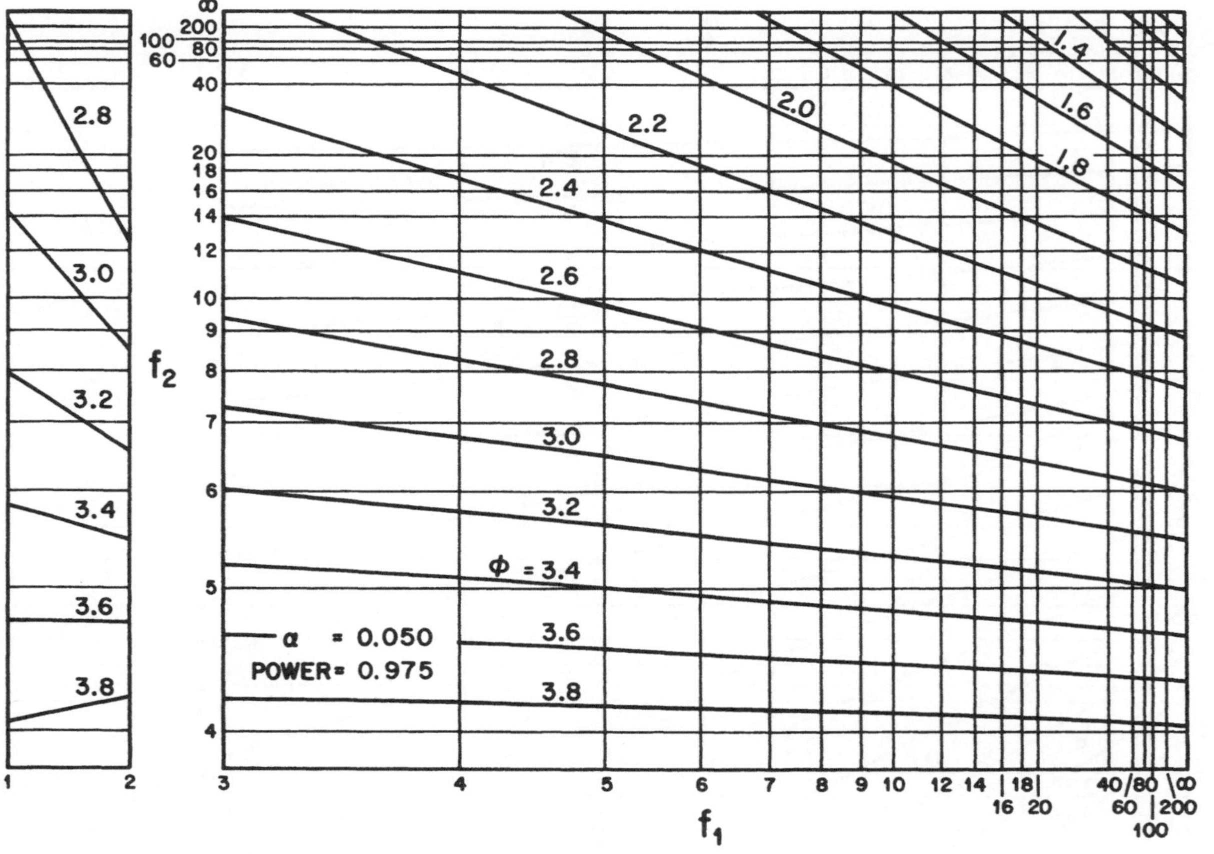

α = 0.050
POWER = 0.975
φ = 3.4
1.4
1.6
1.8
2.0
2.2
2.4
2.6
2.8
3.0
3.2
3.6
3.8
f_1
f_2

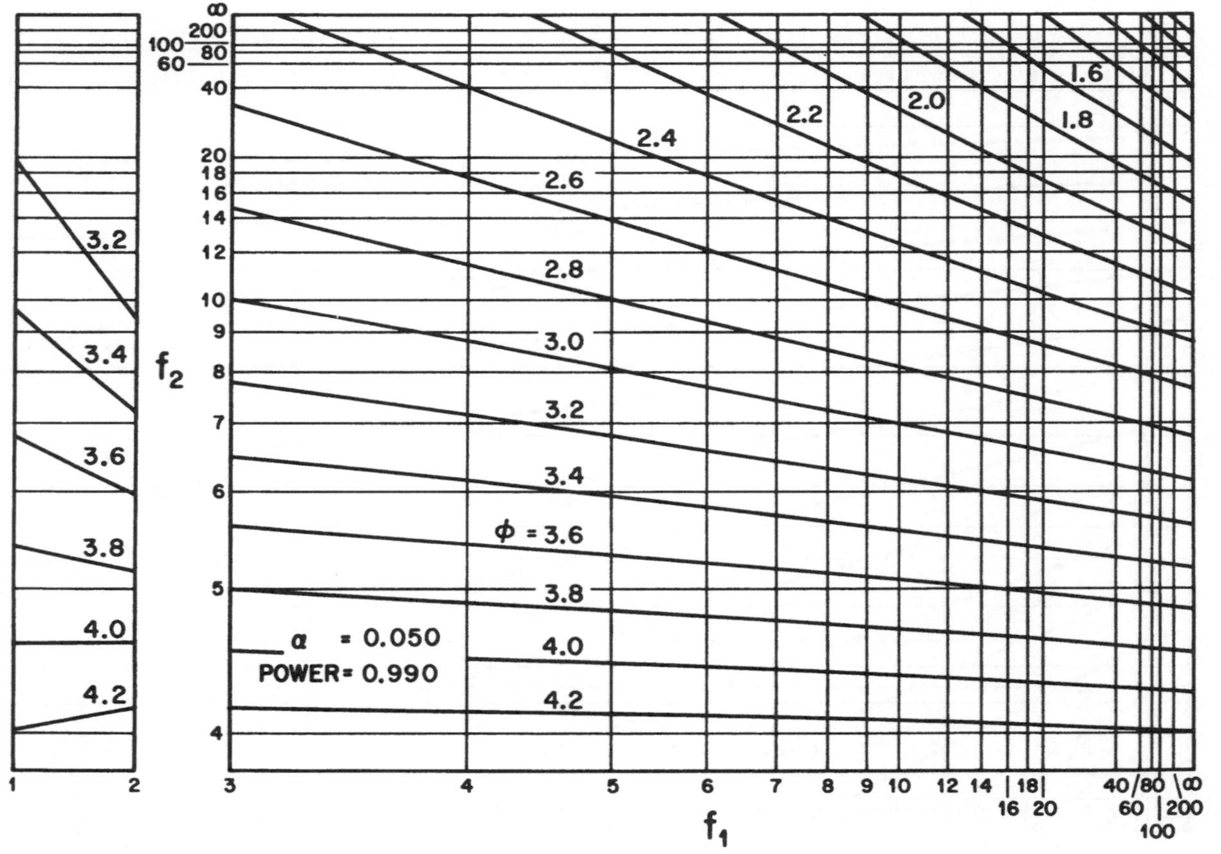
α = 0.050
POWER = 0.990
φ = 3.6
f1
f2
1.6
1.8
2.0
2.2
2.4
2.6
2.8
3.0
3.2
3.4
3.8
4.0
4.2

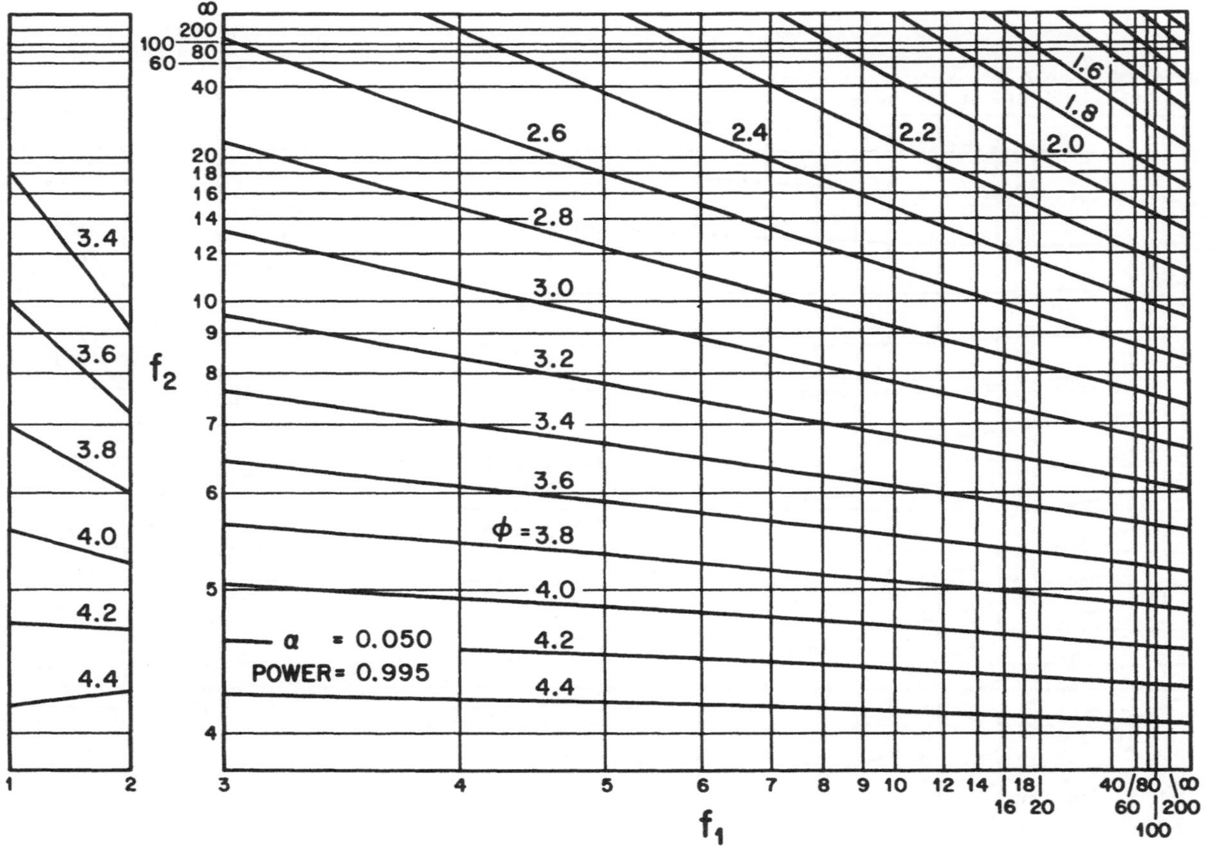
α = 0.050
POWER = 0.995
φ = 3.8
1.6
1.8
2.0
2.2
2.4
2.6
2.8
3.0
3.2
3.4
3.6
4.0
4.2
4.4
3.4
3.6
3.8
4.0
4.2
4.4
f1
f2

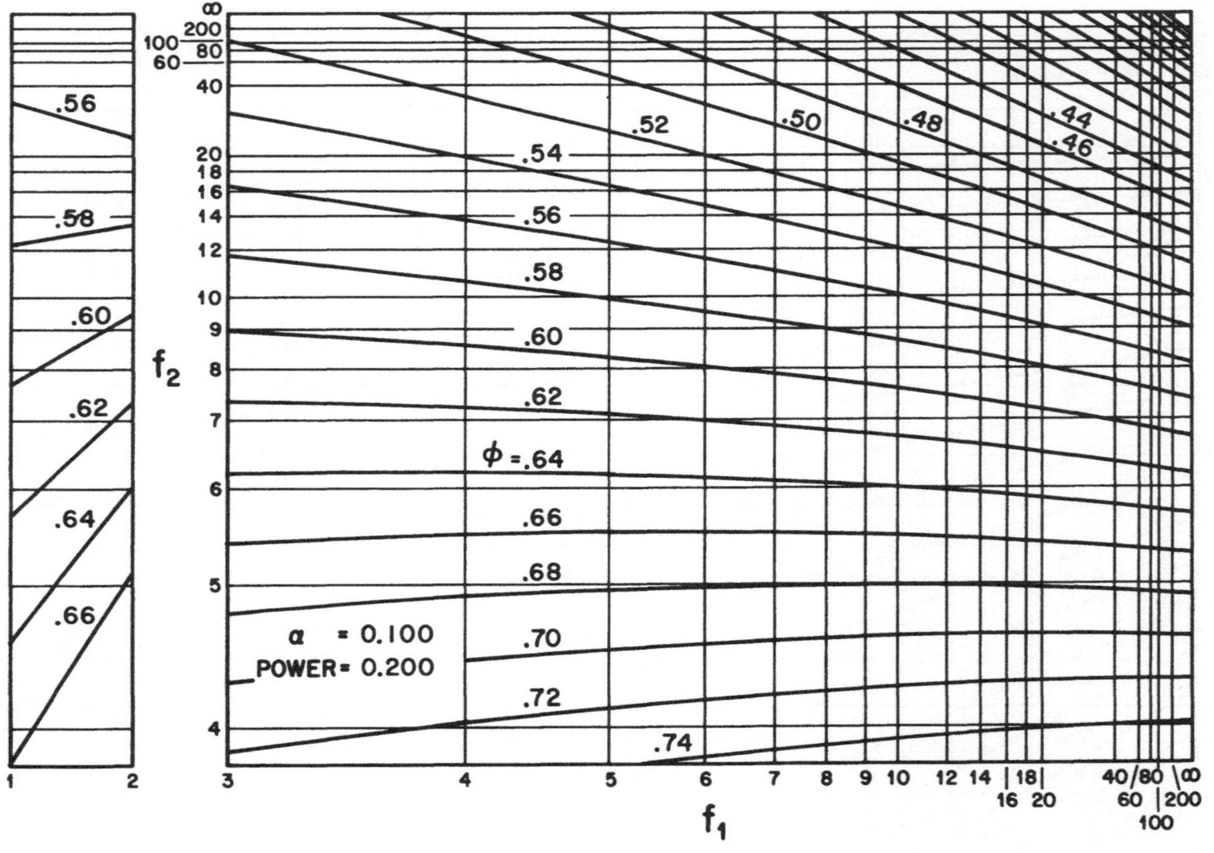
α = 0.100
POWER = 0.200
φ = .64
.44
.46
.48
.50
.52
.54
.56
.58
.60
.62
.66
.68
.70
.72
.74
.56
.58
.60
.62
.64
.66
f_1
f_2

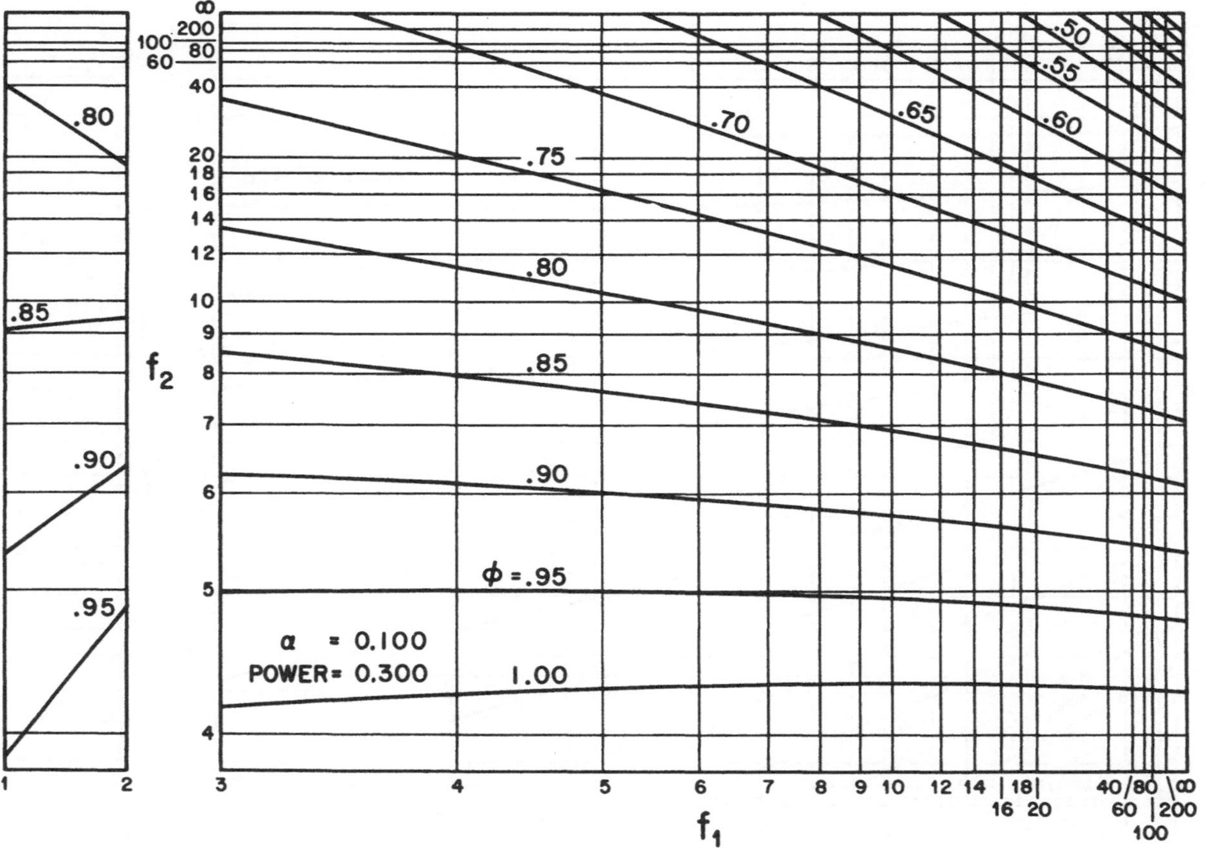

α = 0.100
POWER = 0.300
φ = .95
.50
.55
.60
.65
.70
.75
.80
.85
.90
1.00
.80
.85
.90
.95
f_1
f_2
1
2
3
4
5
6
7
8
9
10
12
14
16
18
20
40
60
80
100
200
∞

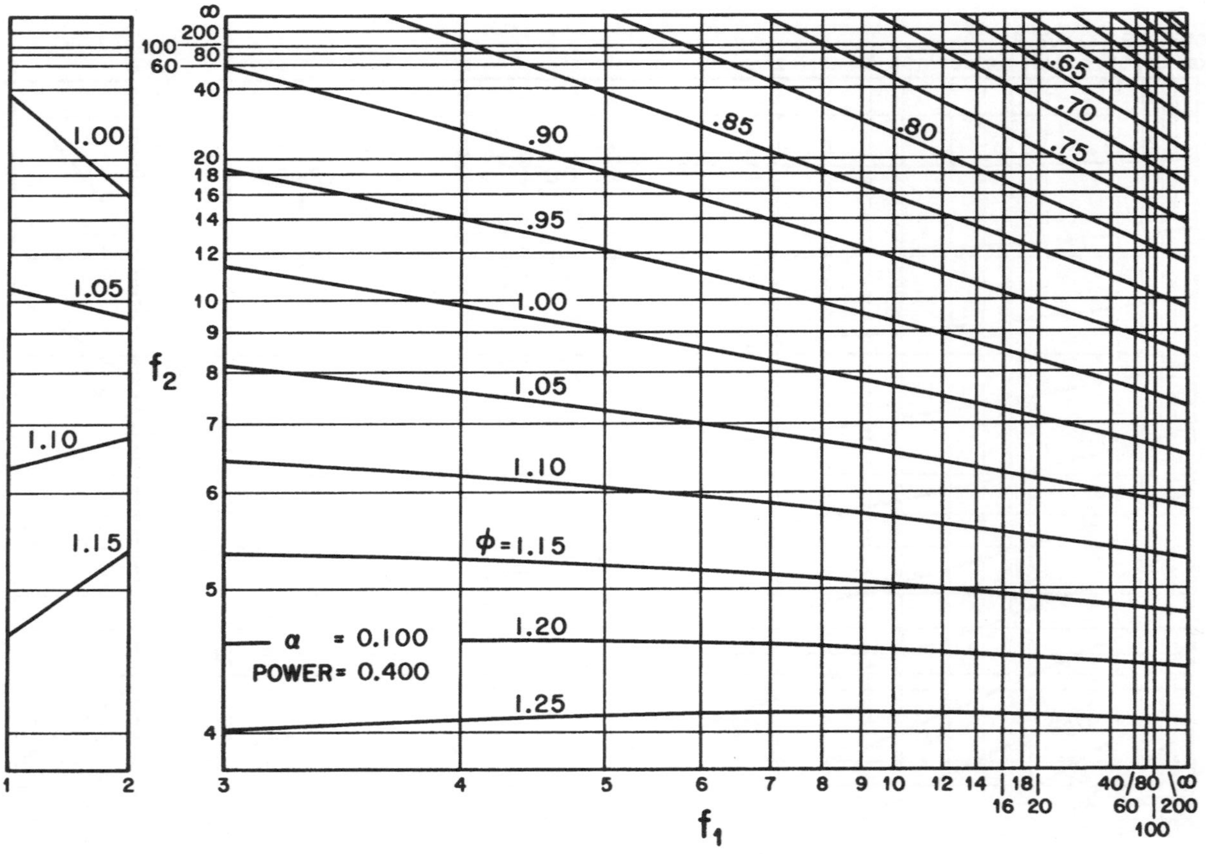
f1
f2
α = 0.100
POWER = 0.400
φ = 1.15
.65
.70
.75
.80
.85
.90
.95
1.00
1.05
1.10
1.20
1.25
1.00
1.05
1.10
1.15

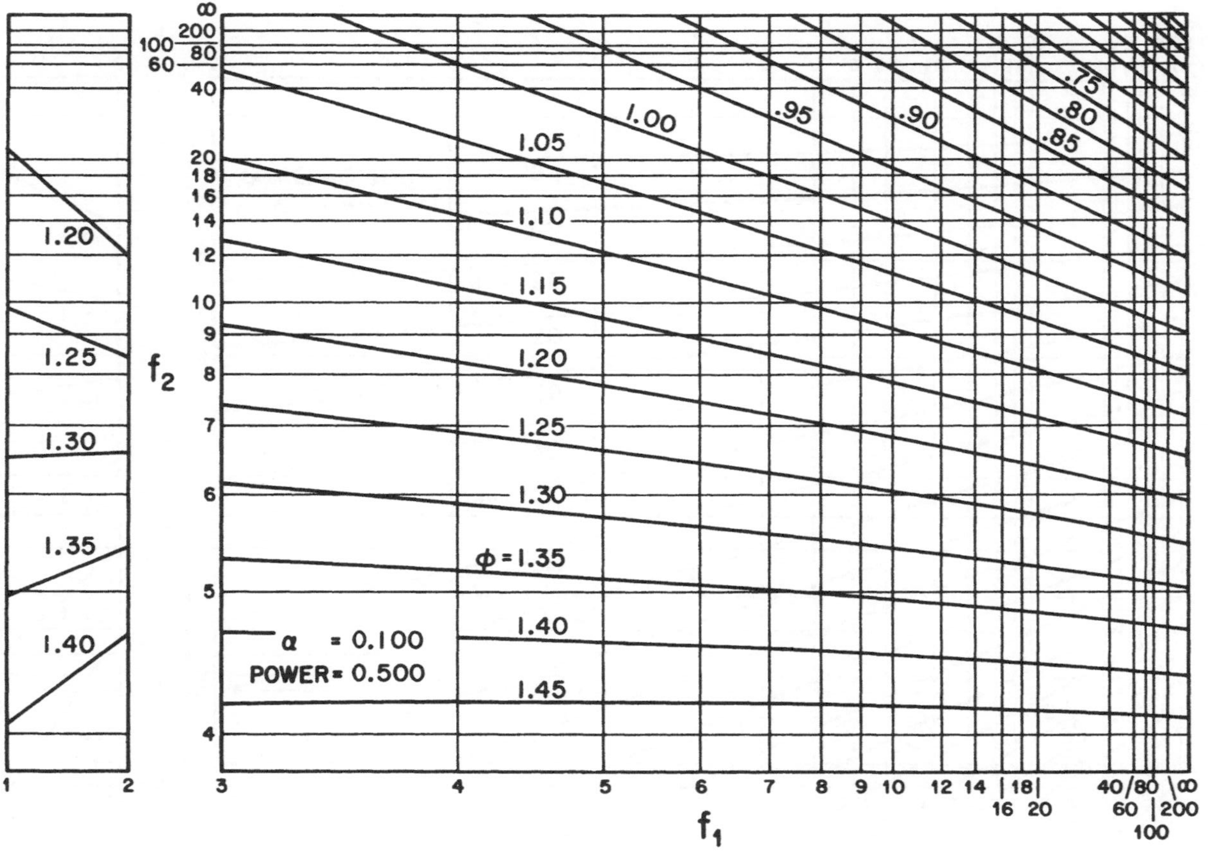

.75
.80
.85
.90
.95
1.00
1.05
1.10
1.15
1.20
1.25
1.30
ϕ = 1.35
1.40
1.45
α = 0.100
POWER = 0.500
f_1
f_2

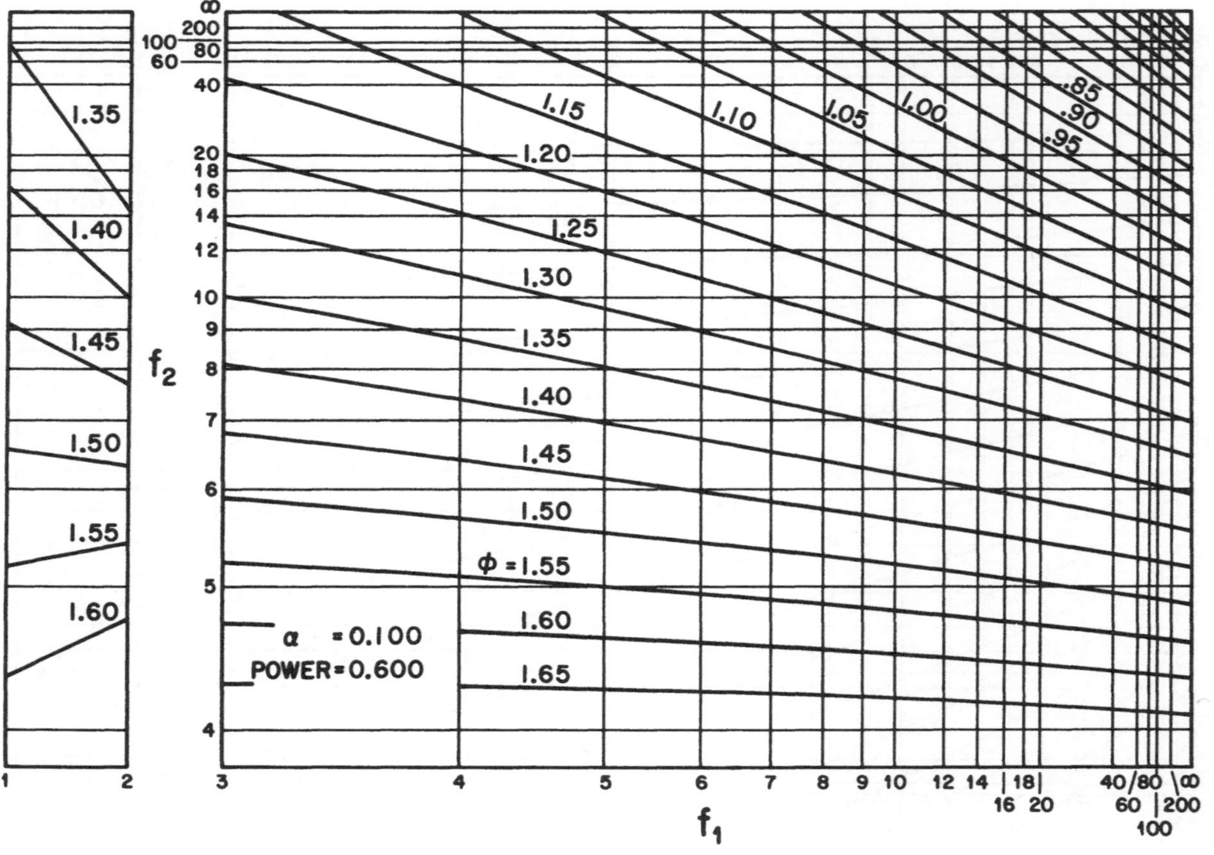

1.35
1.40
1.45
1.50
1.55
1.60
f2
1.15
1.10
1.05
1.00
.85
.90
.95
1.20
1.25
1.30
1.35
1.40
1.45
1.50
ϕ = 1.55
1.60
1.65
α = 0.100
POWER = 0.600
f1

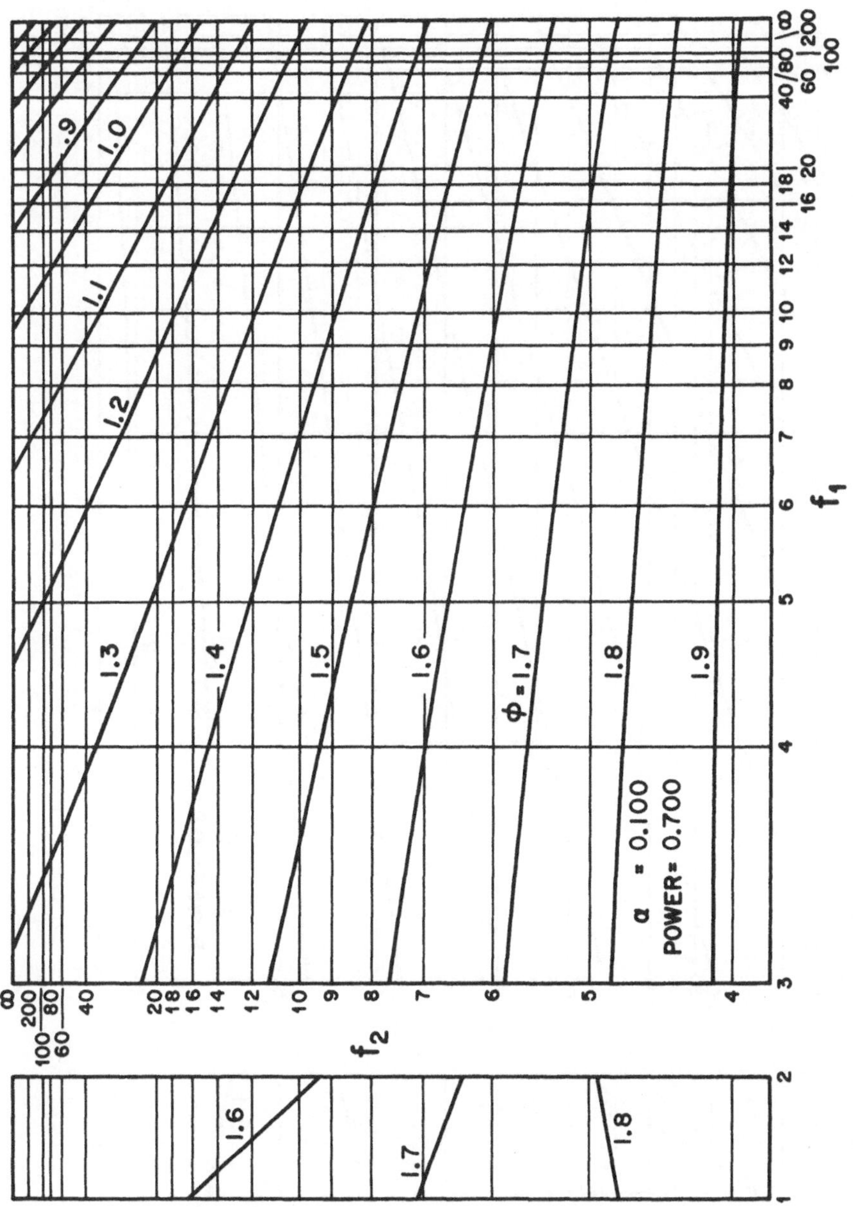

α = 0.100
POWER = 0.700
φ = 1.7
.9
1.0
1.1
1.2
1.3
1.4
1.5
1.6
1.8
1.9
1.6
1.7
1.8
f_1
f_2

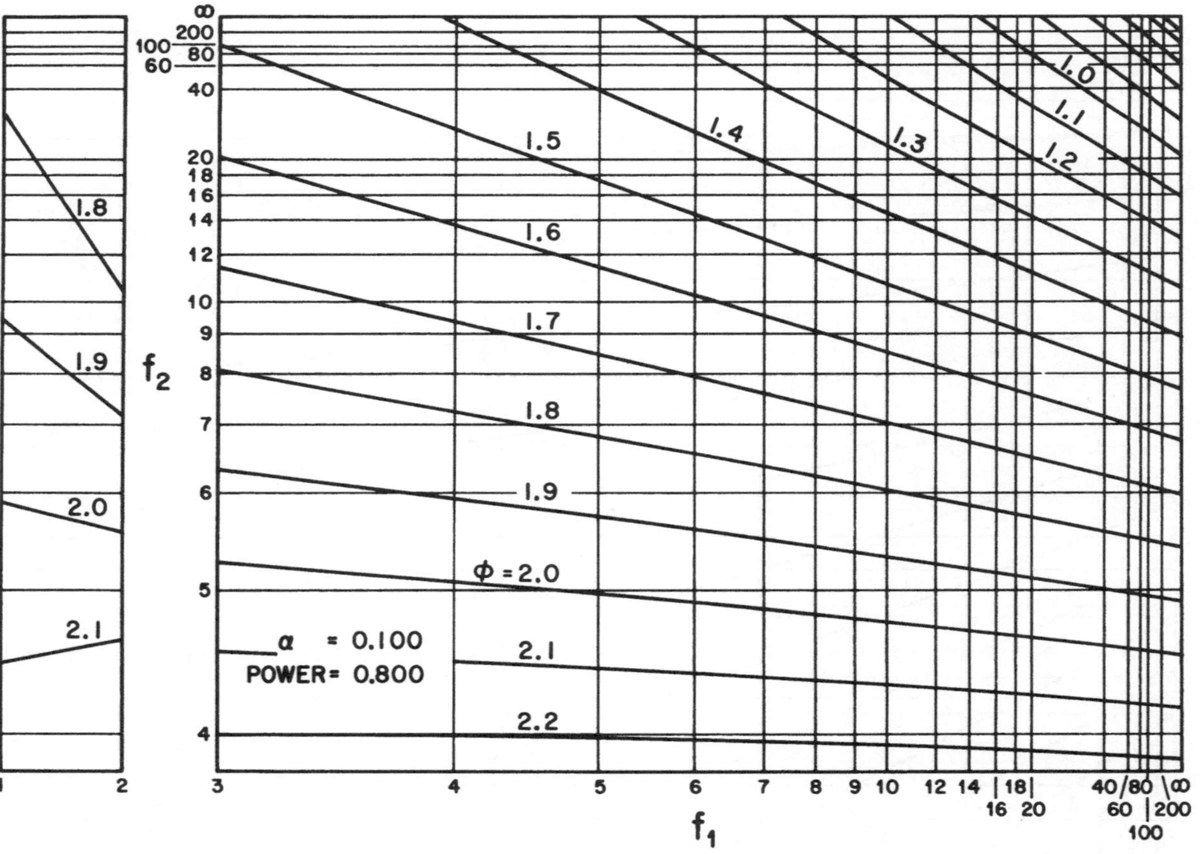
α = 0.100
POWER = 0.800
Φ = 2.0
1.0
1.1
1.2
1.3
1.4
1.5
1.6
1.7
1.8
1.9
2.1
2.2
1.8
1.9
2.0
2.1
f_1
f_2

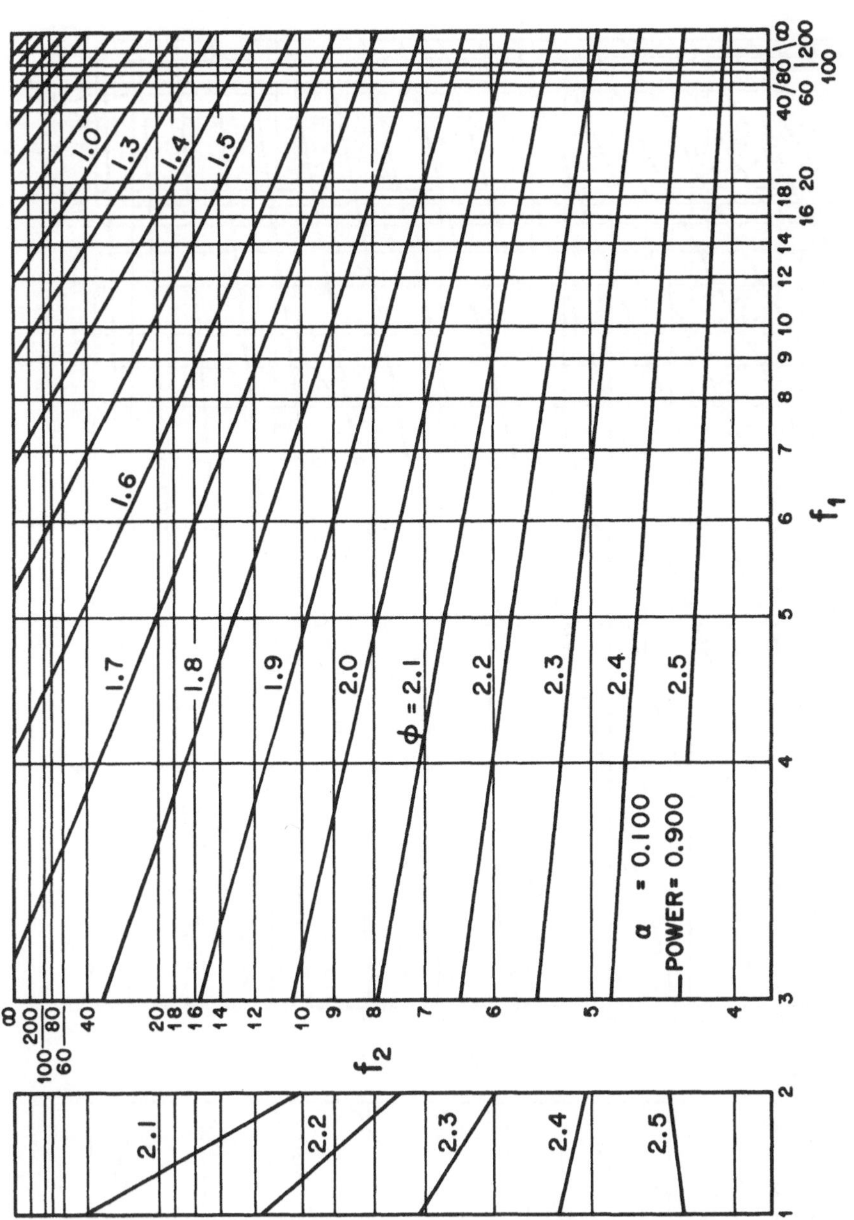

α = 0.100
POWER = 0.900
φ = 2.1
1.0
1.3
1.4
1.5
1.6
1.7
1.8
1.9
2.0
2.2
2.3
2.4
2.5
f_1
f_2

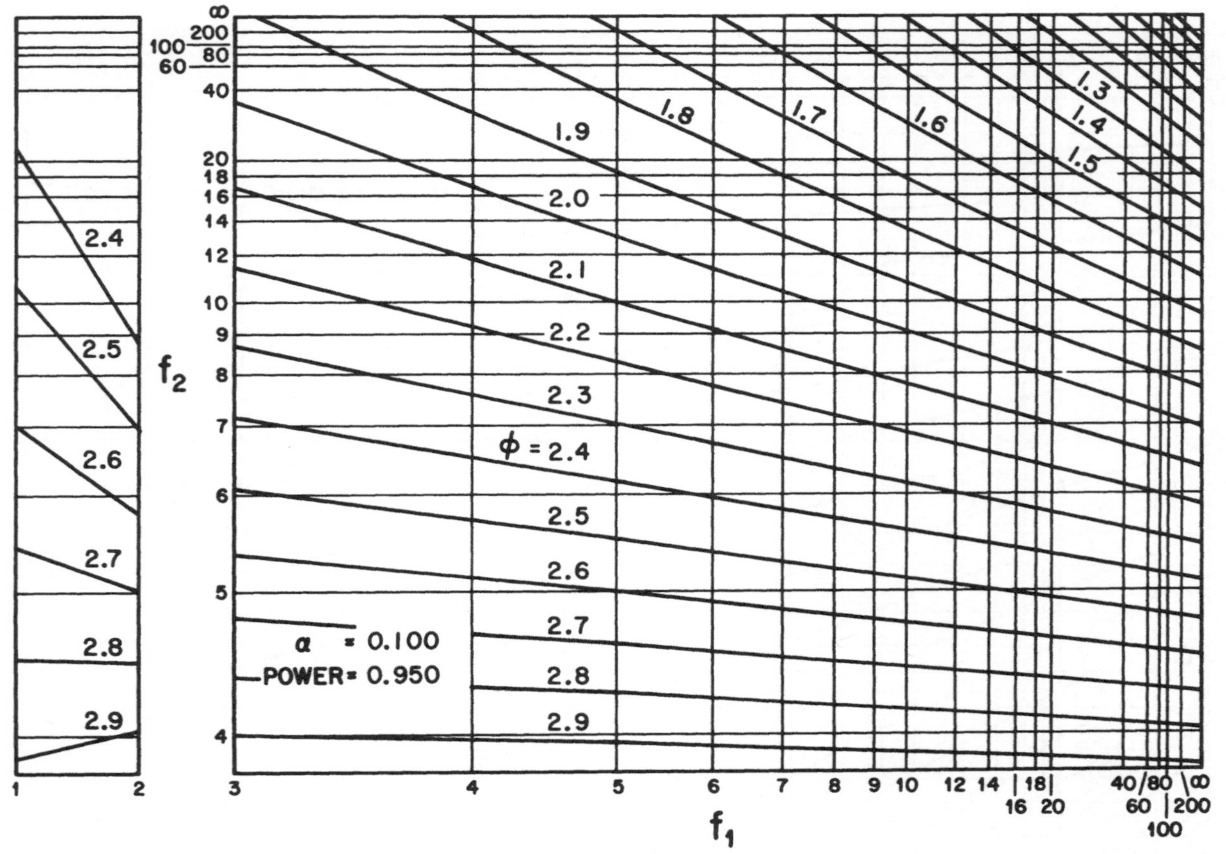

1.3
1.4
1.5
1.6
1.7
1.8
1.9
2.0
2.1
2.2
2.3
φ = 2.4
2.5
2.6
2.7
2.8
2.9
α = 0.100
POWER = 0.950
f_1
f_2

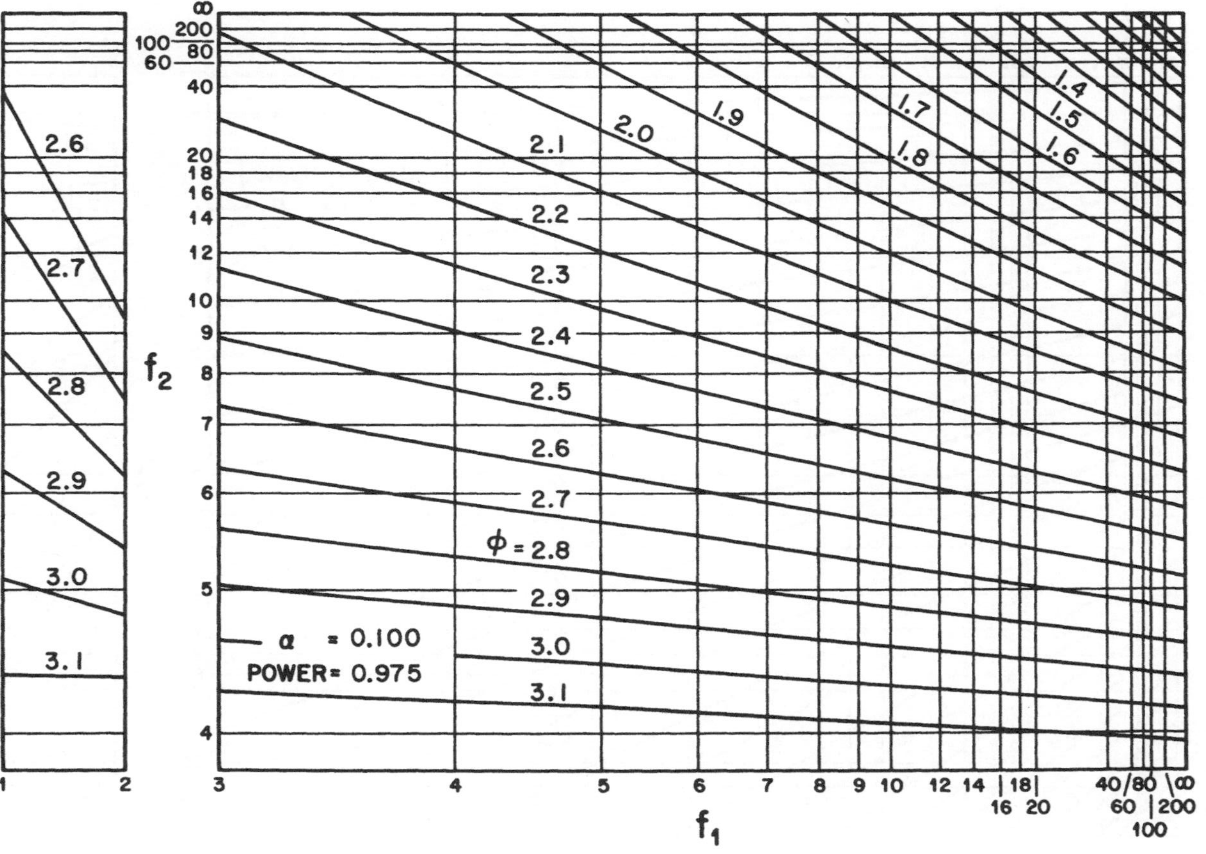
α = 0.100
POWER = 0.975
φ = 2.8
1.4
1.5
1.6
1.7
1.8
1.9
2.0
2.1
2.2
2.3
2.4
2.5
2.6
2.7
2.9
3.0
3.1
f_1
f_2

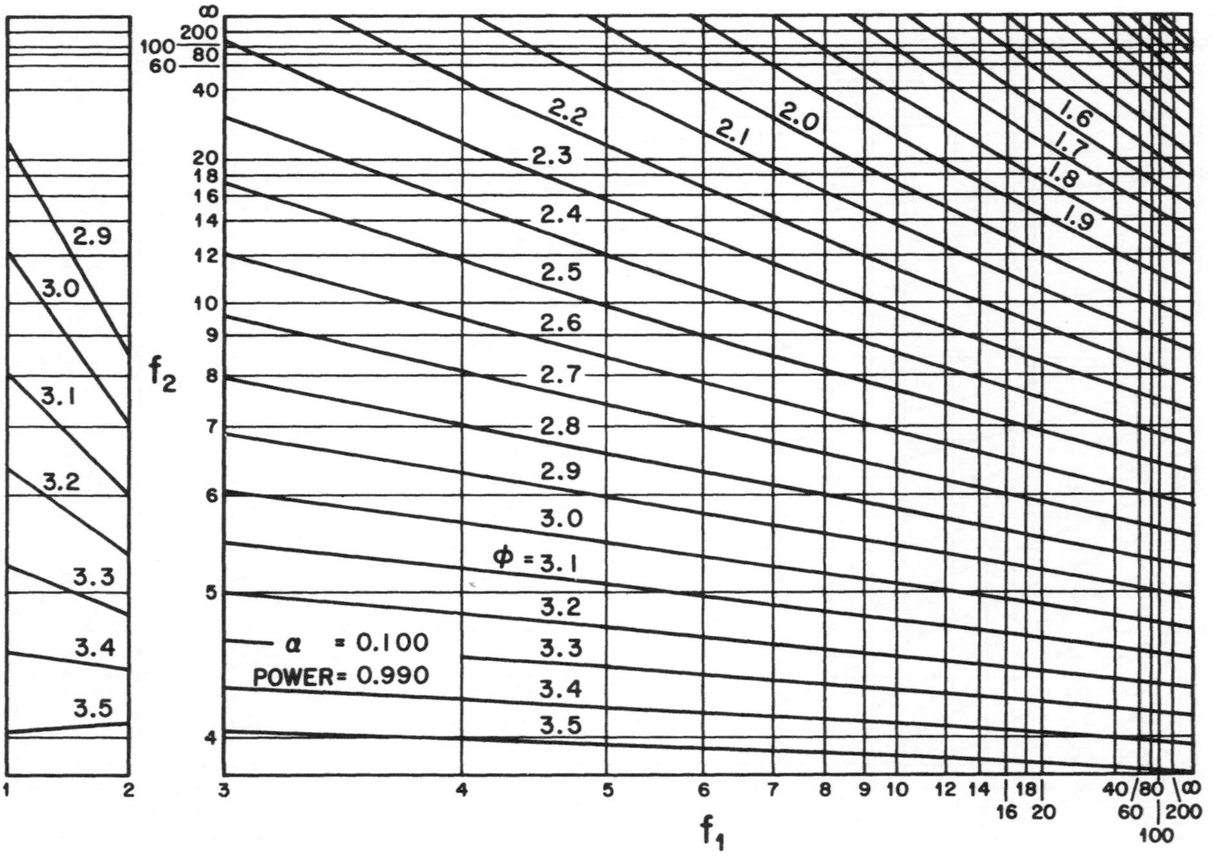
α = 0.100
POWER = 0.990
φ = 3.1
f_1
f_2

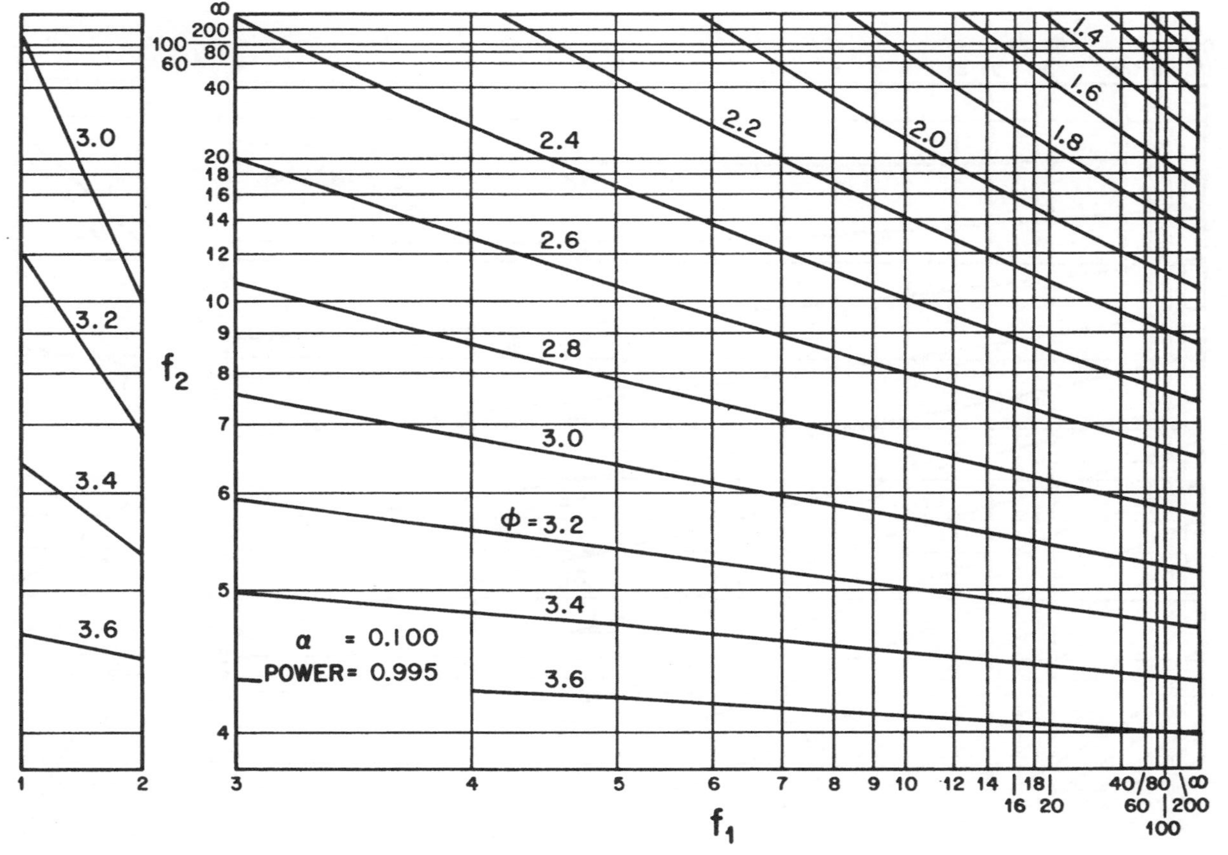
3.0
3.2
3.4
3.6
f_2
2.4
2.2
2.0
1.4
1.6
1.8
2.6
2.8
3.0
φ = 3.2
3.4
α = 0.100
POWER = 0.995
3.6
f_1

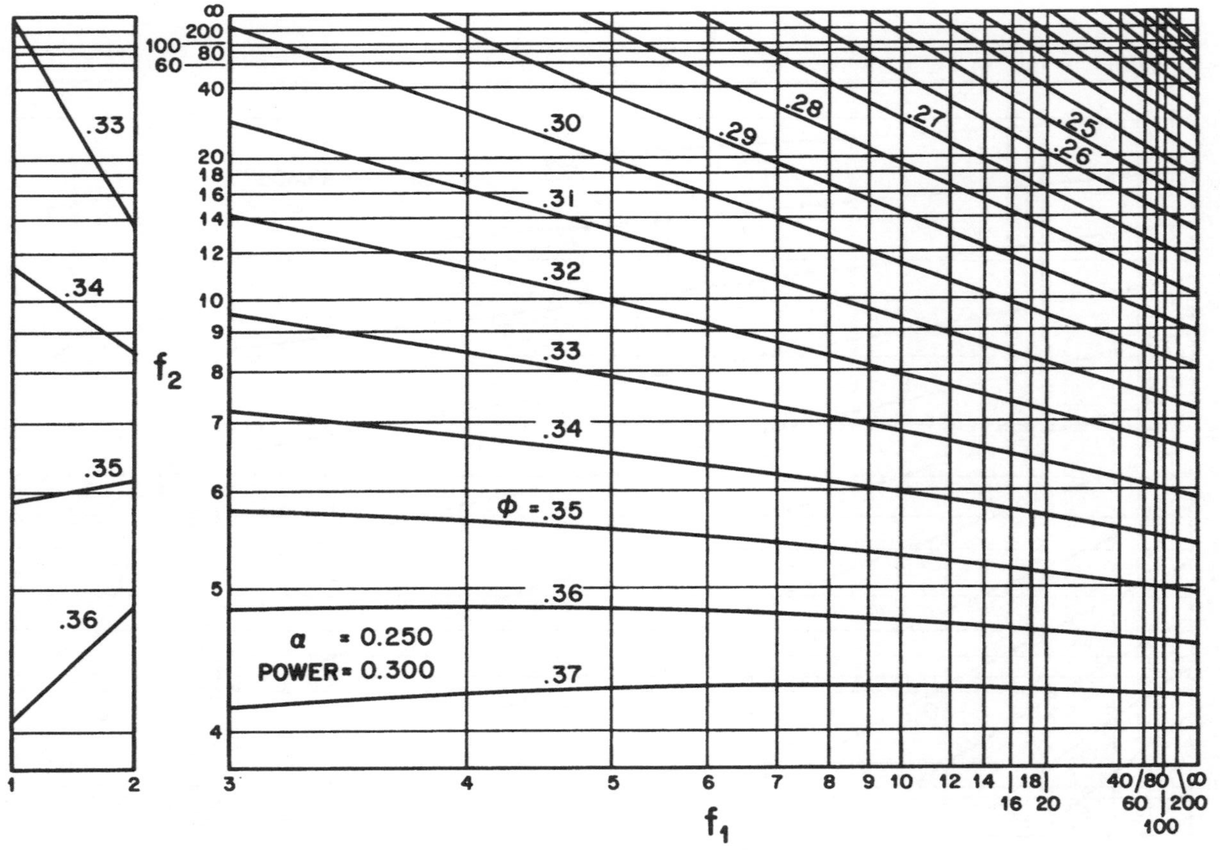
.33
.34
.35
.36
f_2
.30
.29
.28
.27
.25
.26
.31
.32
.33
.34
φ = .35
.36
α = 0.250
POWER = 0.300
.37
f_1

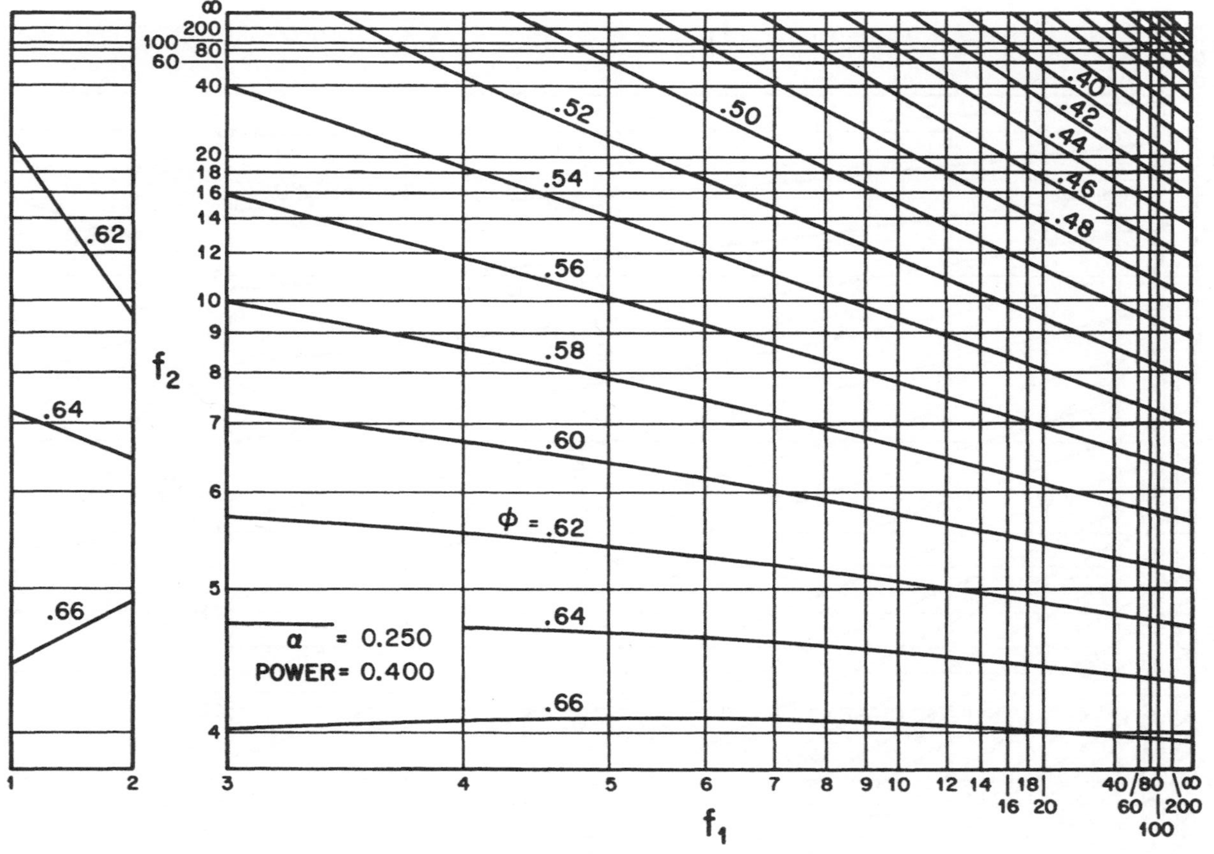

α = 0.250
POWER = 0.400
φ = .62
.40
.42
.44
.46
.48
.50
.52
.54
.56
.58
.60
.64
.66
f1
f2

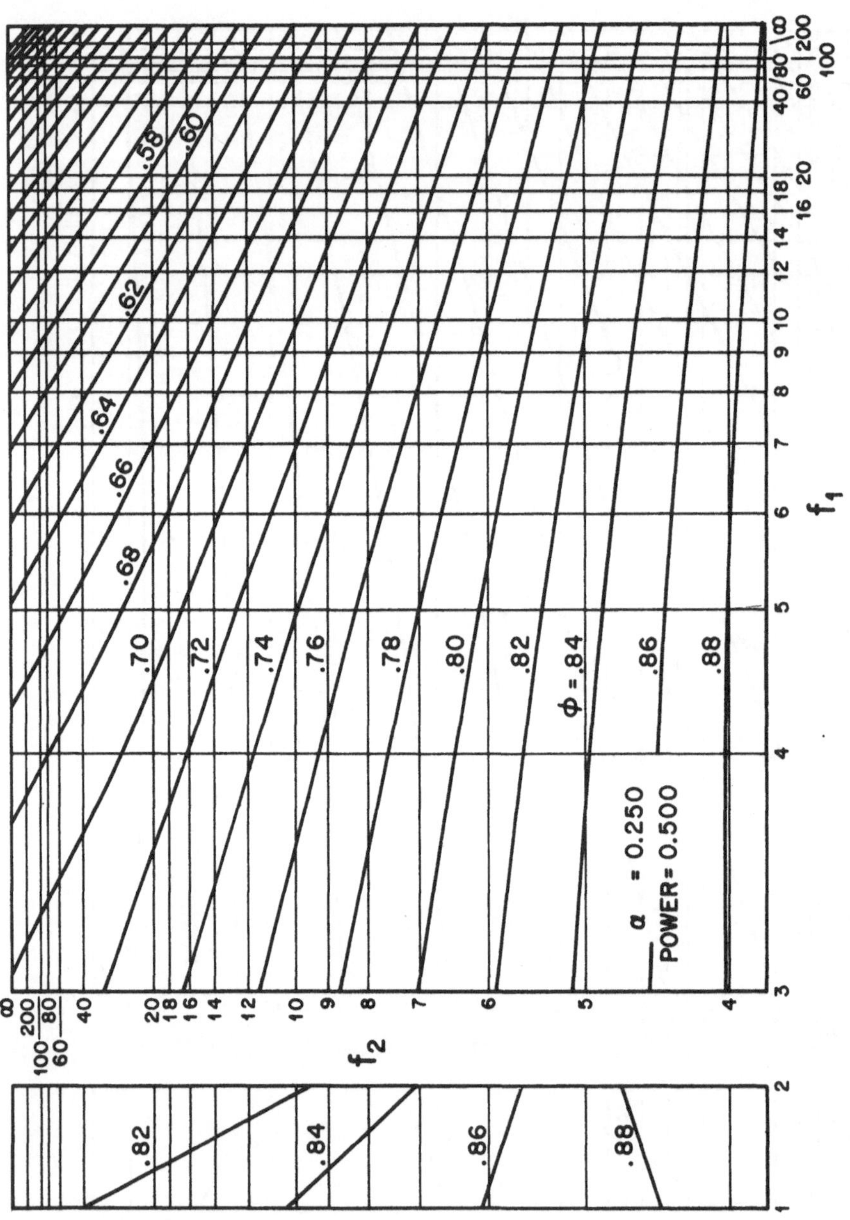

α = 0.250
POWER = 0.500
φ = .84
.58
.60
.62
.64
.66
.68
.70
.72
.74
.76
.78
.80
.82
.86
.88
.82
.84
.86
.88
f_1
f_2

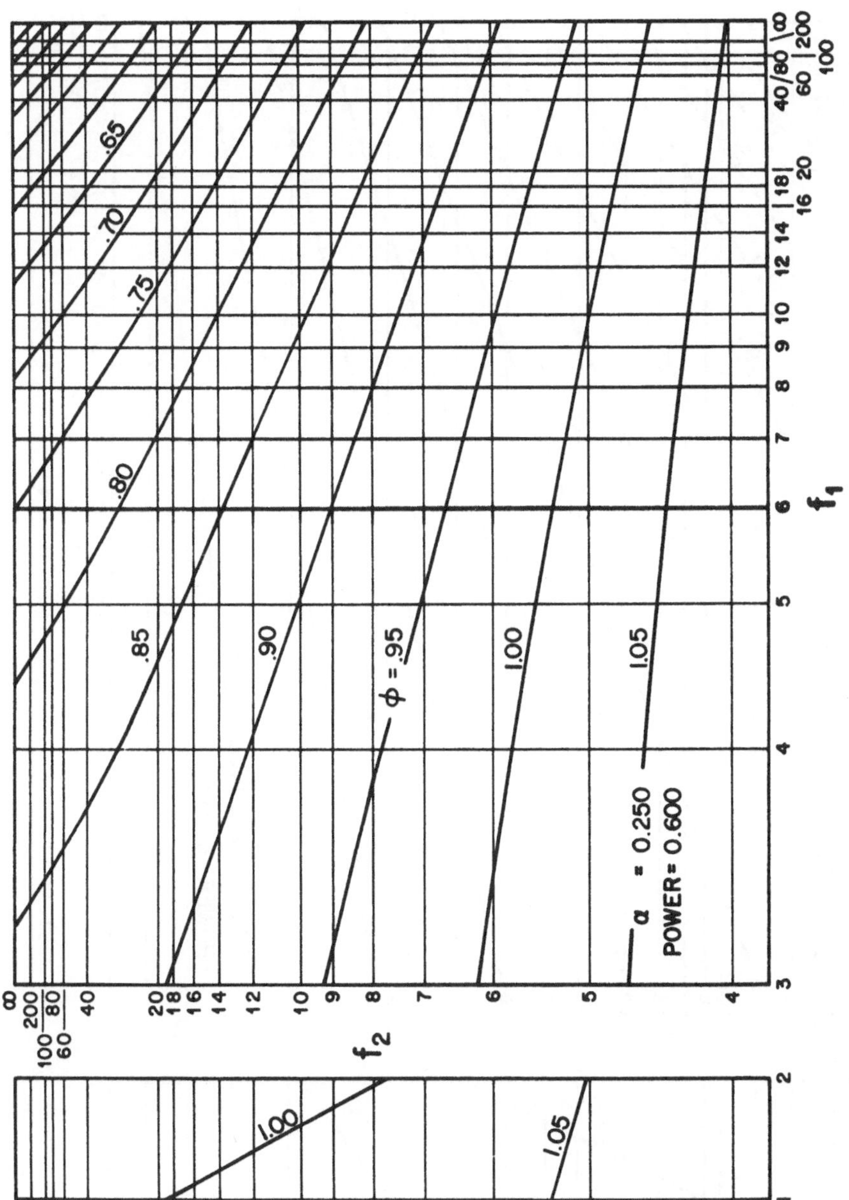

α = 0.250
POWER = 0.600
φ = .95
.65
.70
.75
.80
.85
.90
1.00
1.05
f_1
f_2

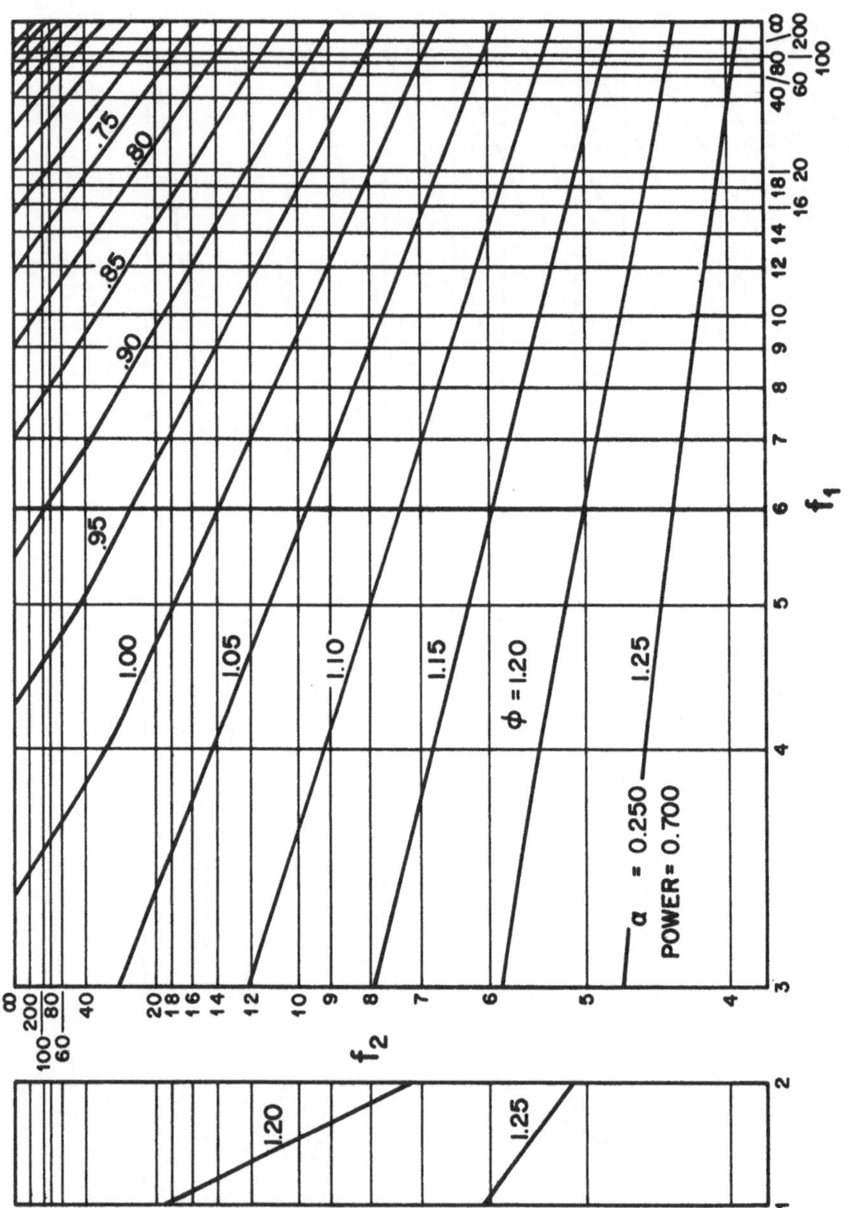

f_1
f_2
.75
.80
.85
.90
.95
1.00
1.05
1.10
1.15
ϕ = 1.20
1.25
1.20
α = 0.250
POWER = 0.700

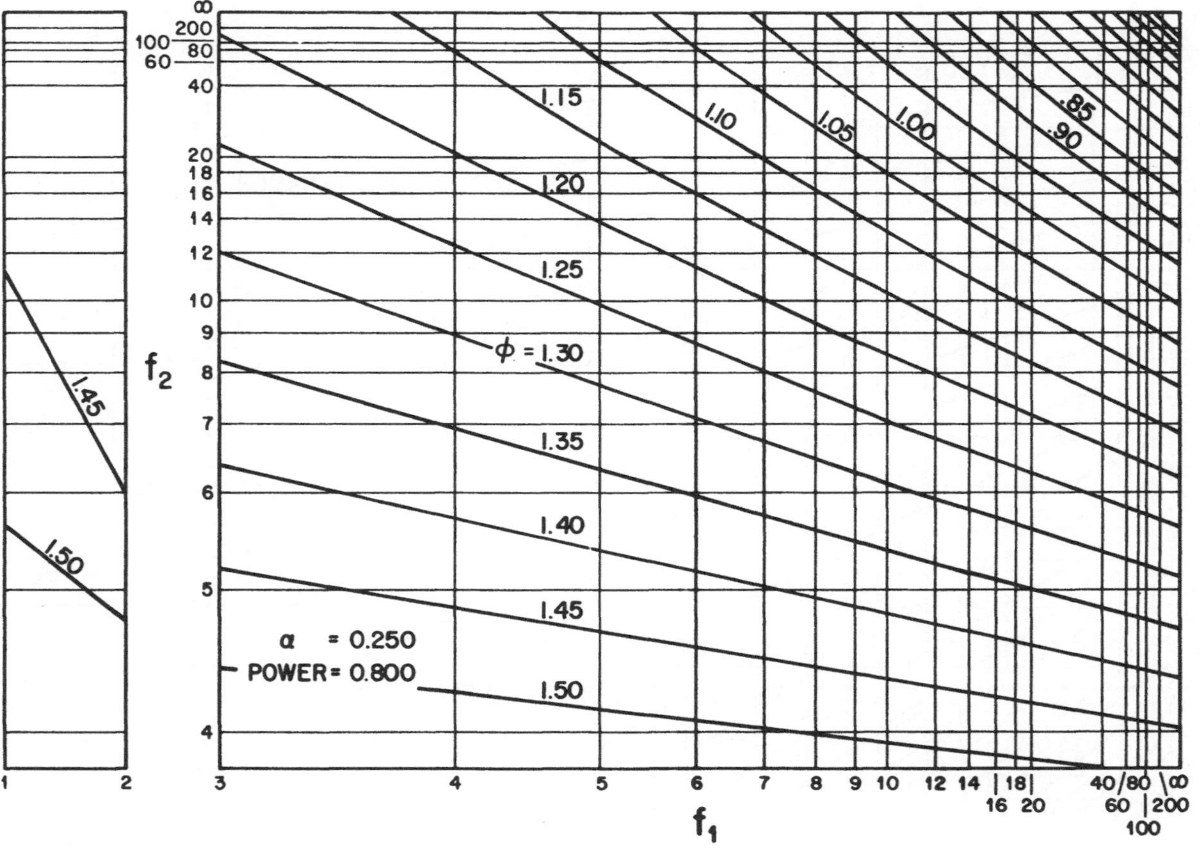

α = 0.250
POWER = 0.800
φ = 1.30
.85
.90
1.00
1.05
1.10
1.15
1.20
1.25
1.35
1.40
1.45
1.50
f1
f2

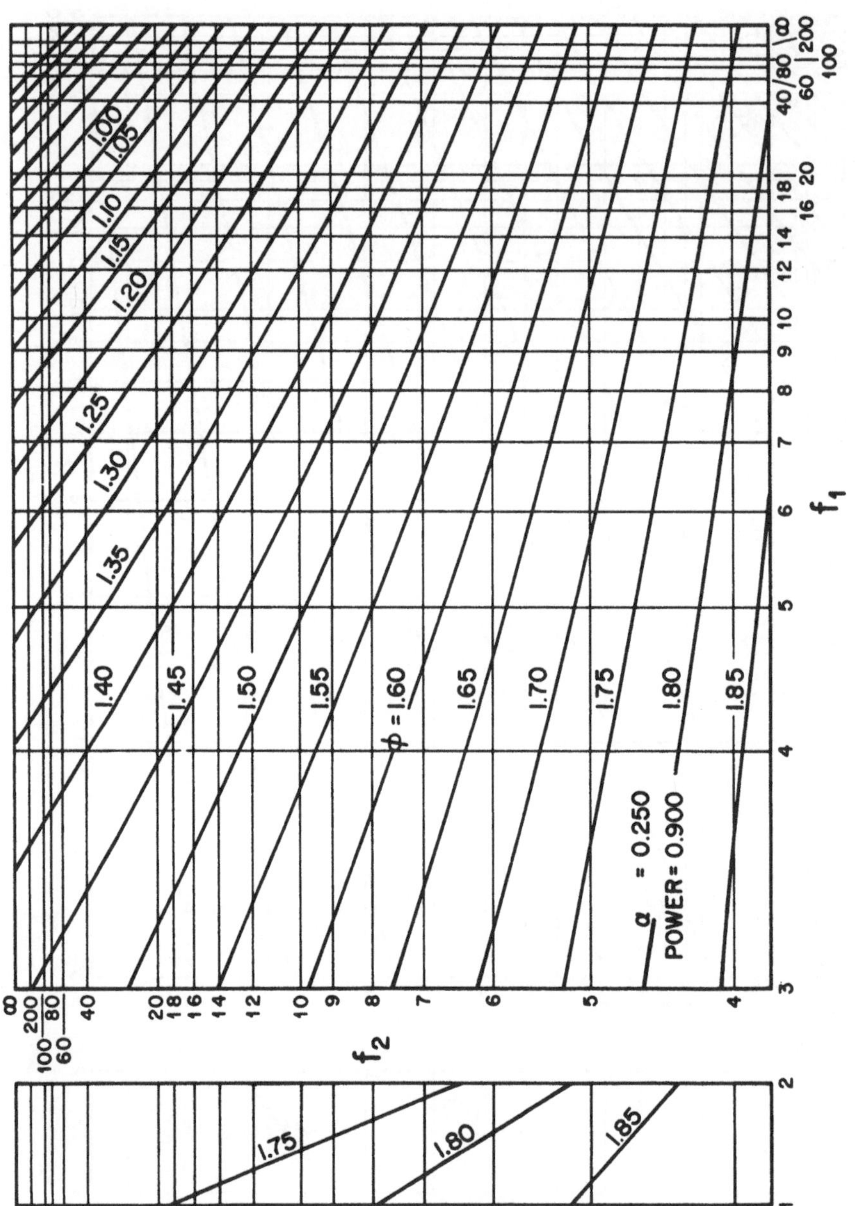

α = 0.250
POWER = 0.900
φ = 1.60
1.00
1.05
1.10
1.15
1.20
1.25
1.30
1.35
1.40
1.45
1.50
1.55
1.65
1.70
1.75
1.80
1.85
f_1
f_2

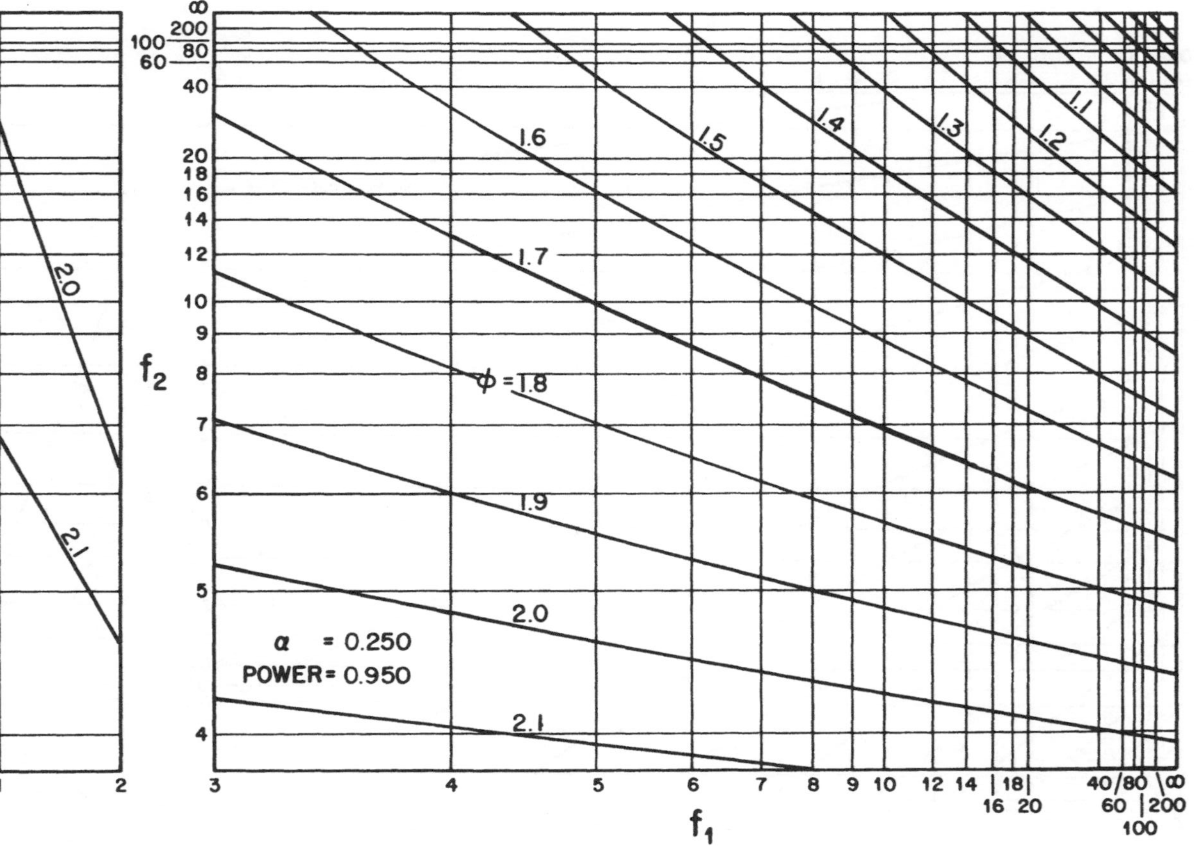
α = 0.250
POWER = 0.950
φ = 1.8
1.1
1.2
1.3
1.4
1.5
1.6
1.7
1.9
2.0
2.1
f_1
f_2
1
2
3
4
5
6
7
8
9
10
12
14
16
18
20
40
60
80
100
200
∞

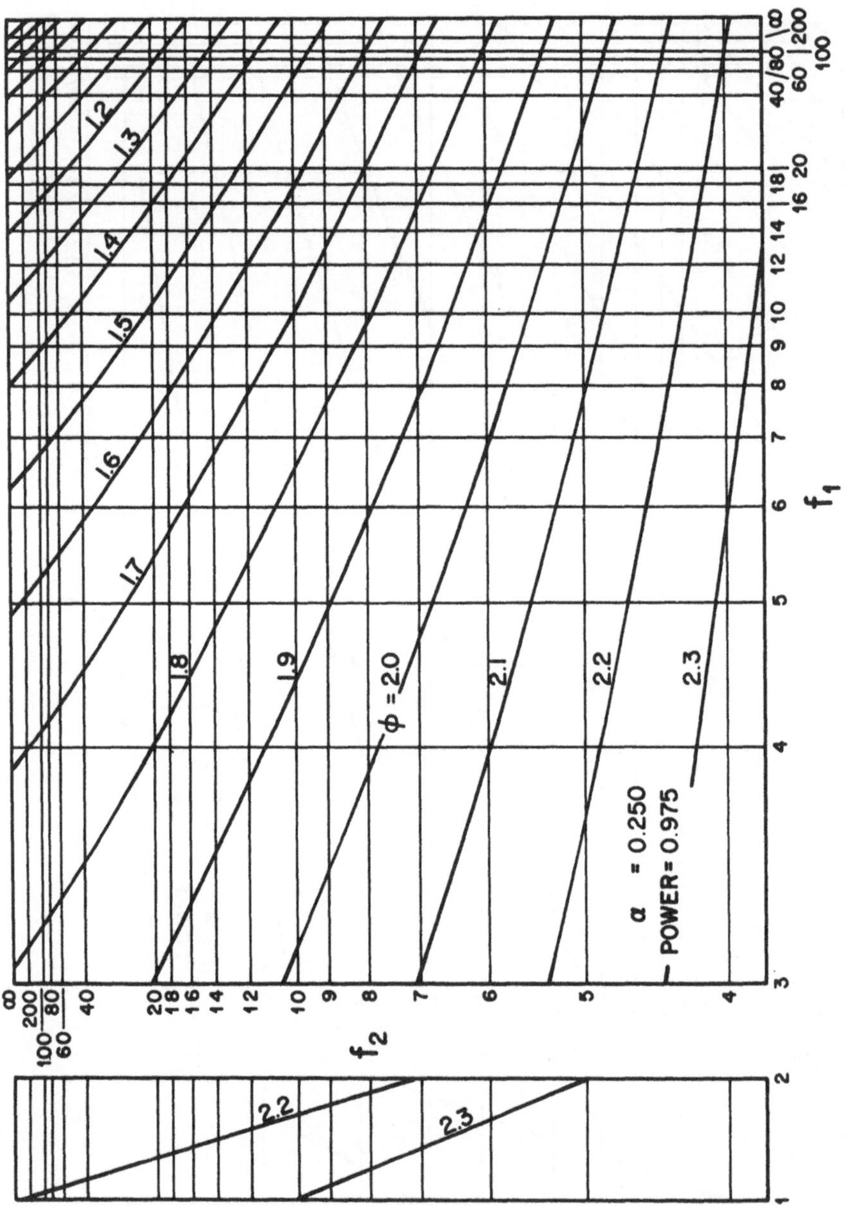
α = 0.250
POWER = 0.975
φ = 2.0
1.2
1.3
1.4
1.5
1.6
1.7
1.8
1.9
2.1
2.2
2.3
f_1
f_2

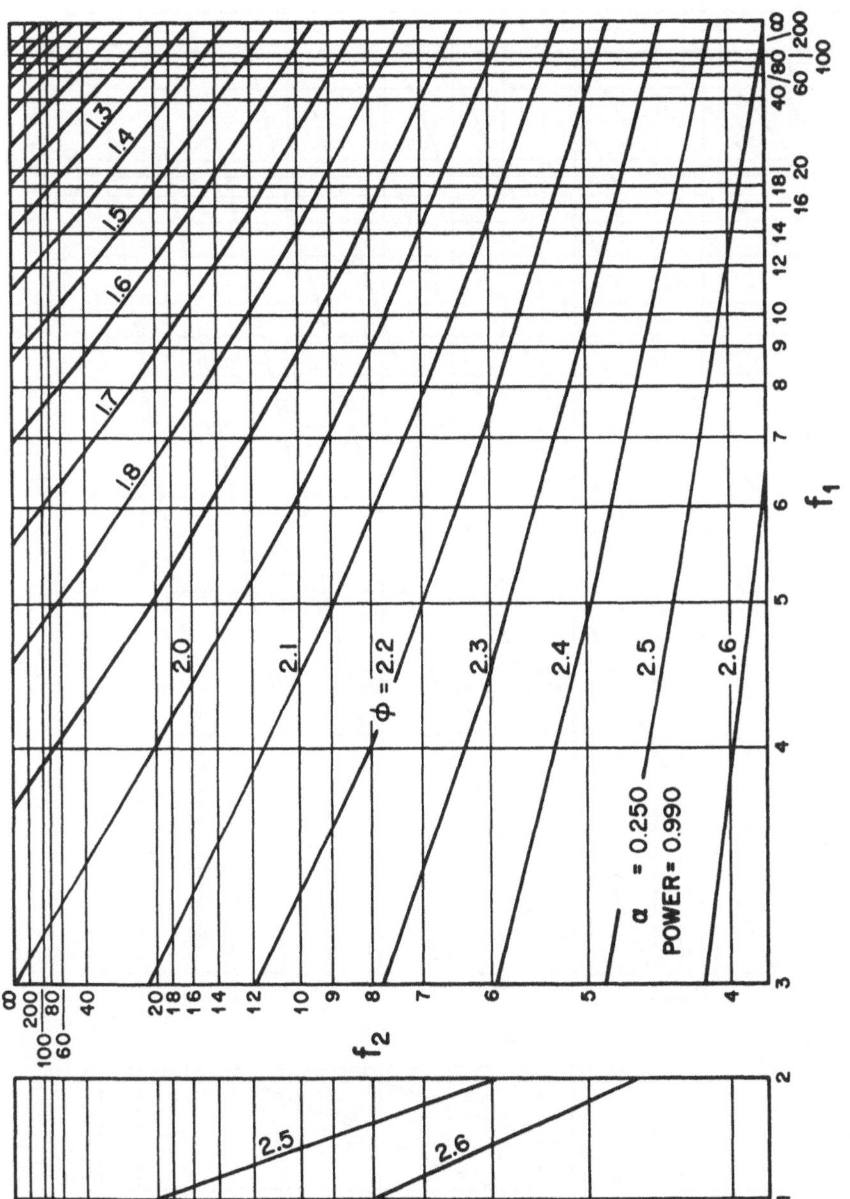
α = 0.250
POWER = 0.990
φ = 2.2
1.3
1.4
1.5
1.6
1.7
1.8
2.0
2.1
2.3
2.4
2.5
2.6
f_1
f_2

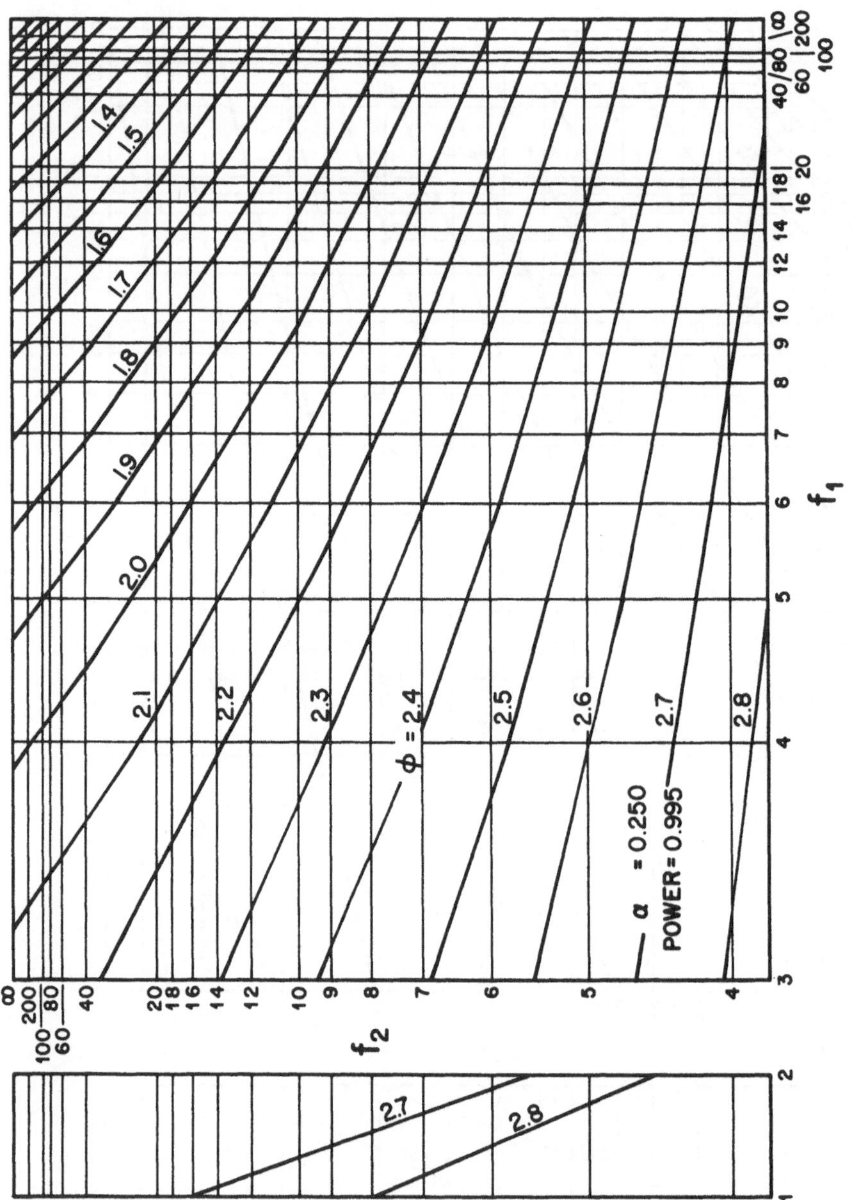

α = 0.250
POWER = 0.995
φ = 2.4
f_1
f_2

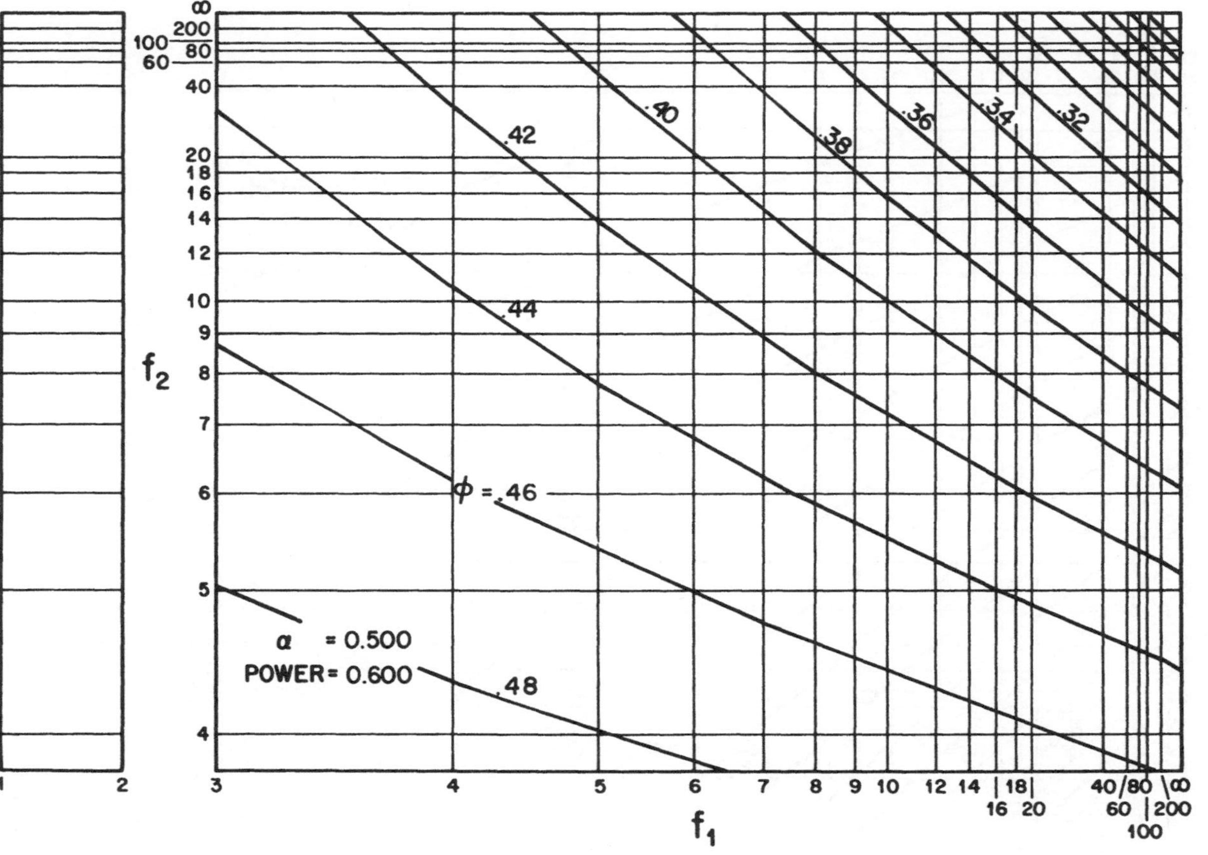

α = 0.500
POWER = 0.600
φ = .46
.48
.44
.42
.40
.38
.36
.34
.32
f_1
f_2
1 2 3 4 5 6 7 8 9 10 12 14 16 18 20 40/80 60 100 200 ∞
4 5 6 7 8 9 10 12 14 16 18 20 40 60 80 100 200 ∞

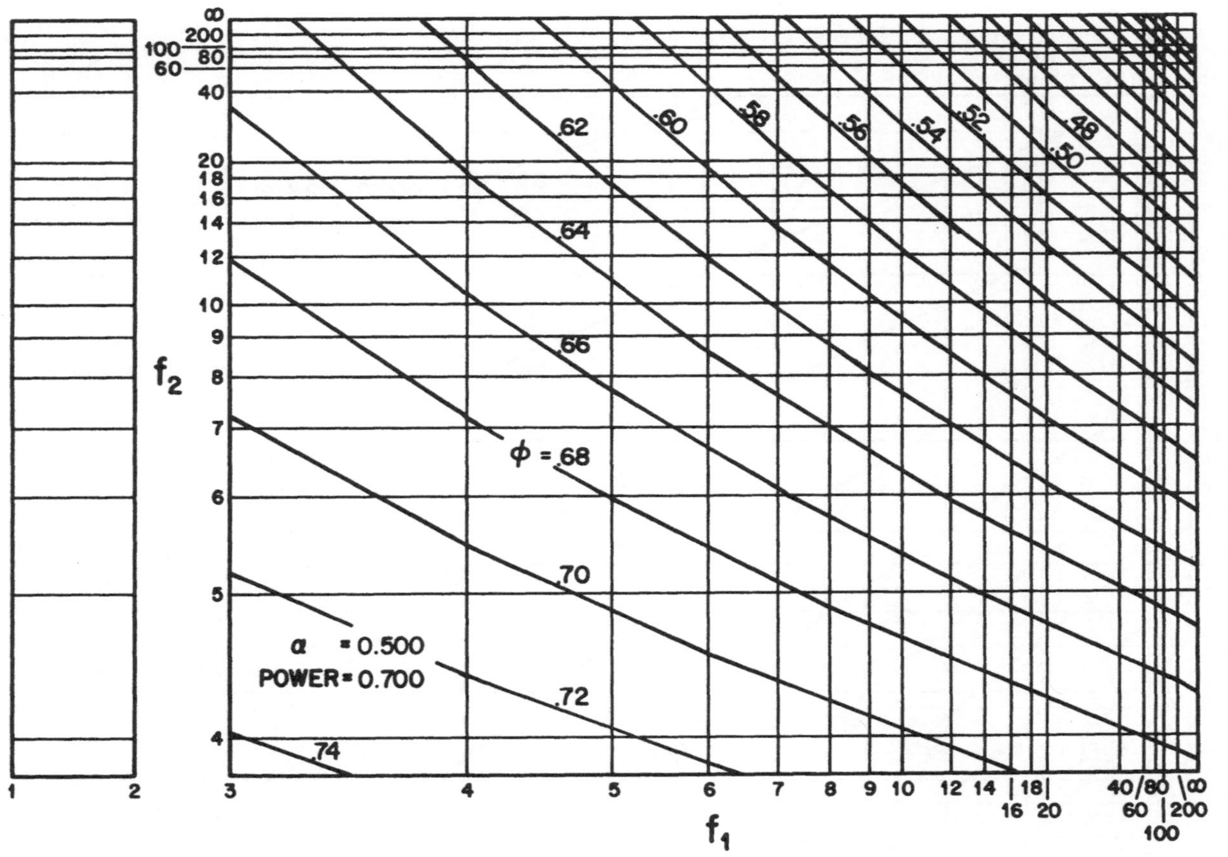

α = 0.500
POWER = 0.700
ϕ = .68
.48
.50
.52
.54
.56
.58
.60
.62
.64
.66
.70
.72
.74
f_1
f_2

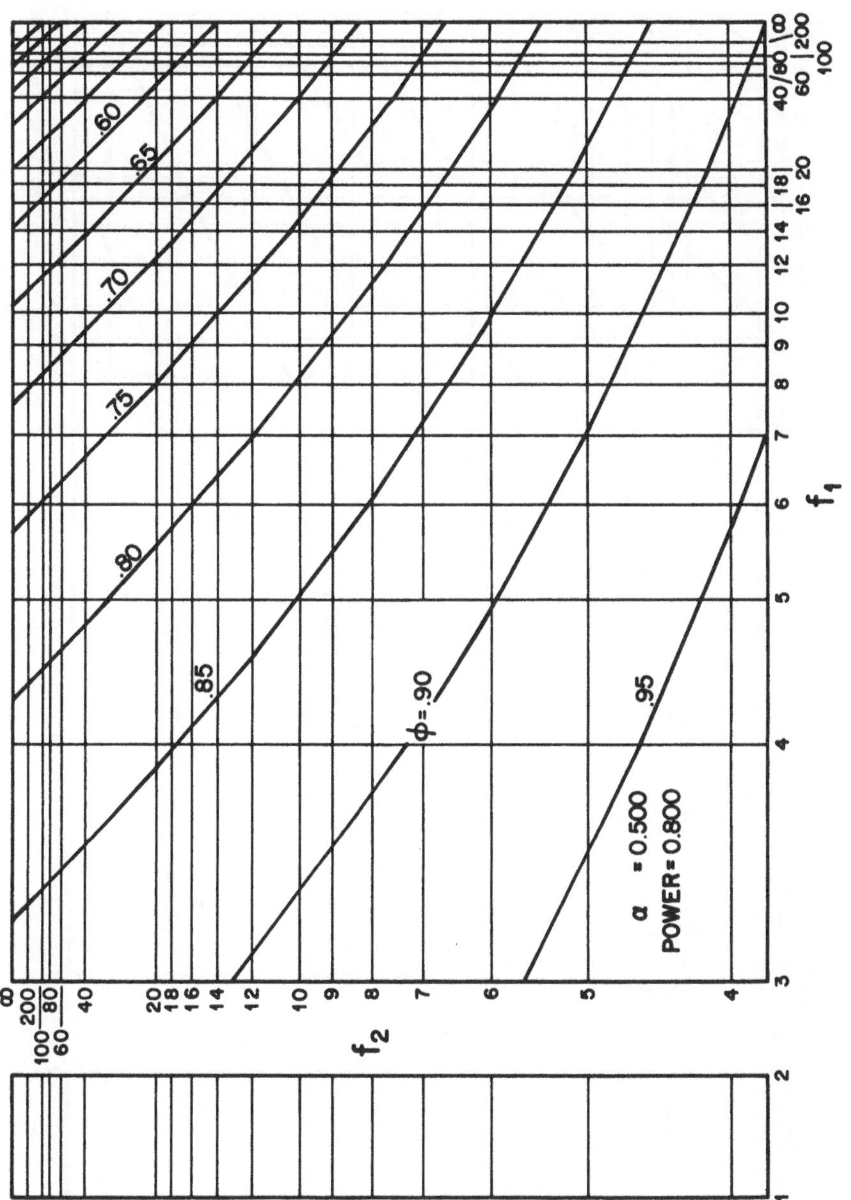

α = 0.500
POWER = 0.800
φ = .90
.95
.85
.80
.75
.70
.65
.60
f1
f2

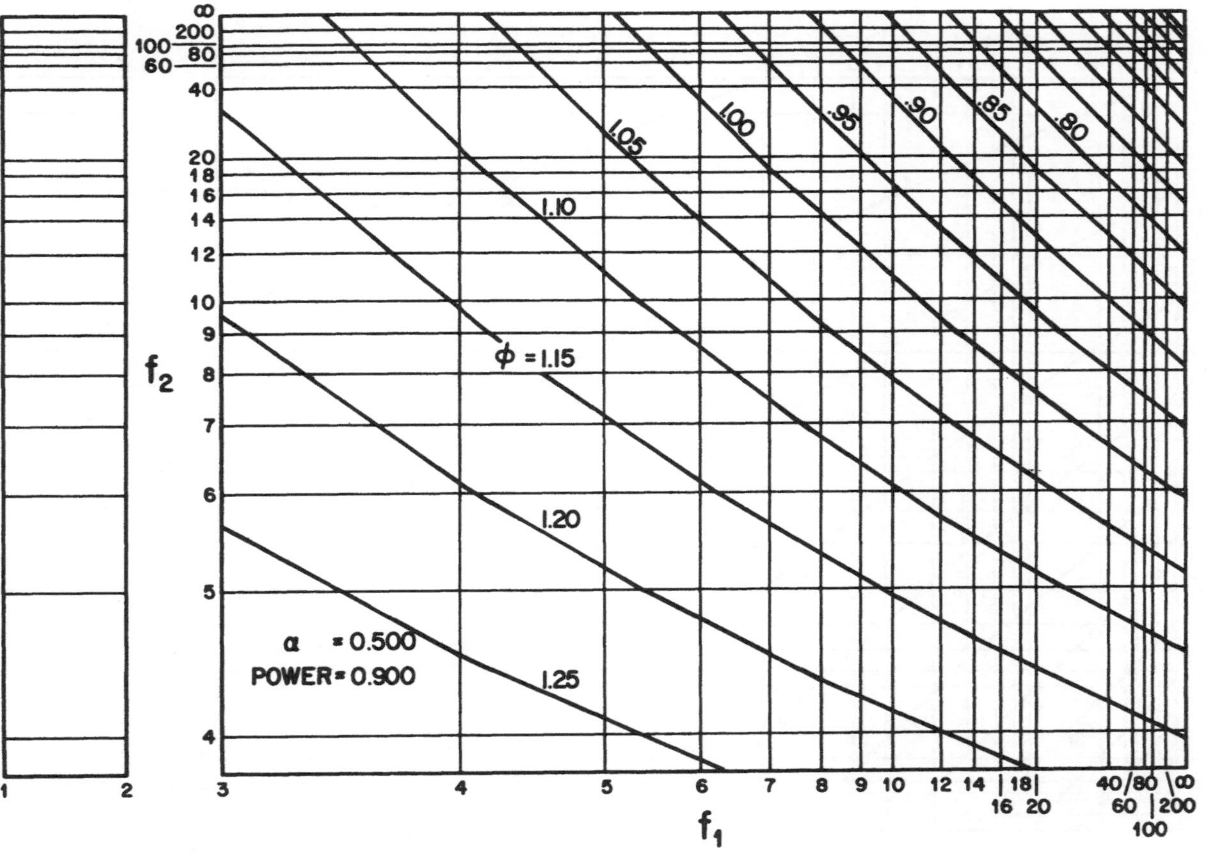
α = 0.500
POWER = 0.900
φ = 1.15
.80
.85
.90
.95
1.00
1.05
1.10
1.20
1.25
f_1
f_2

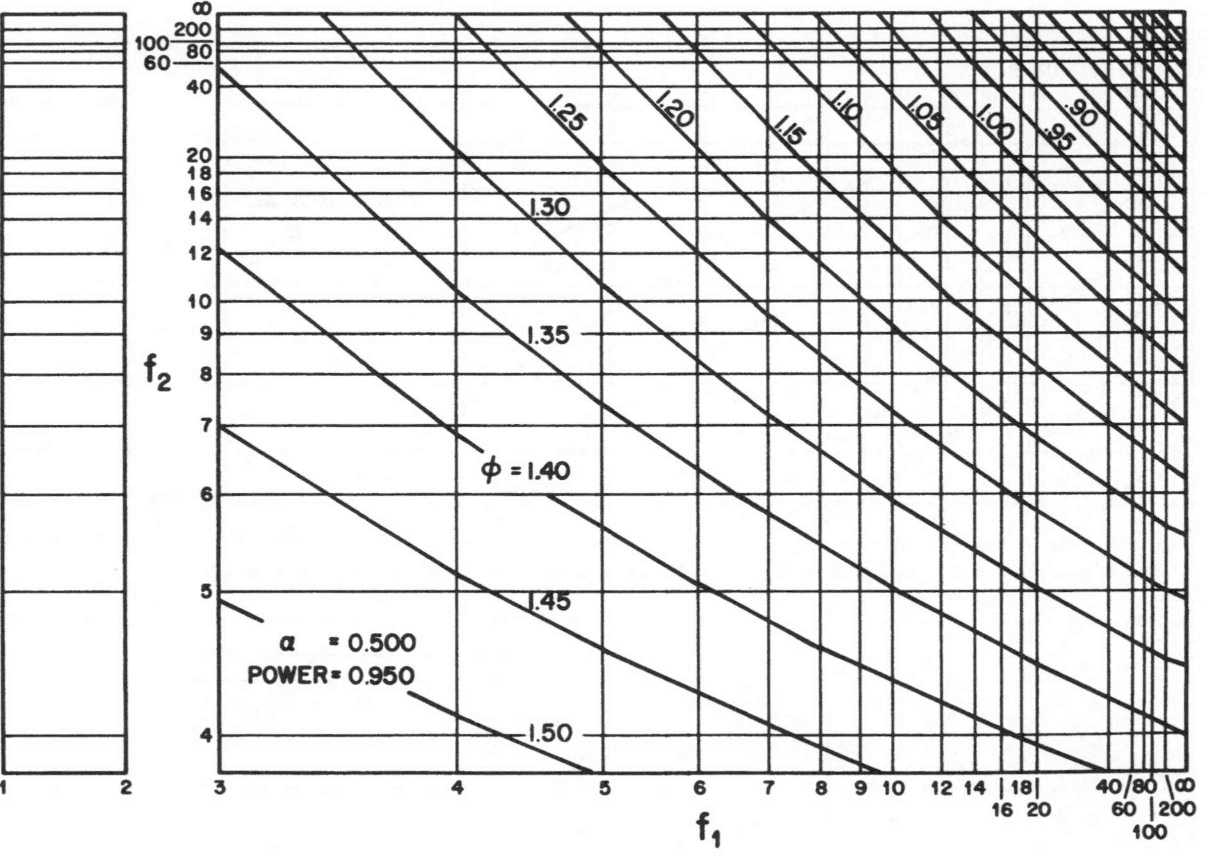

α = 0.500
POWER = 0.950
φ = 1.40
.90
.95
1.00
1.05
1.10
1.15
1.20
1.25
1.30
1.35
1.45
1.50
f1
f2

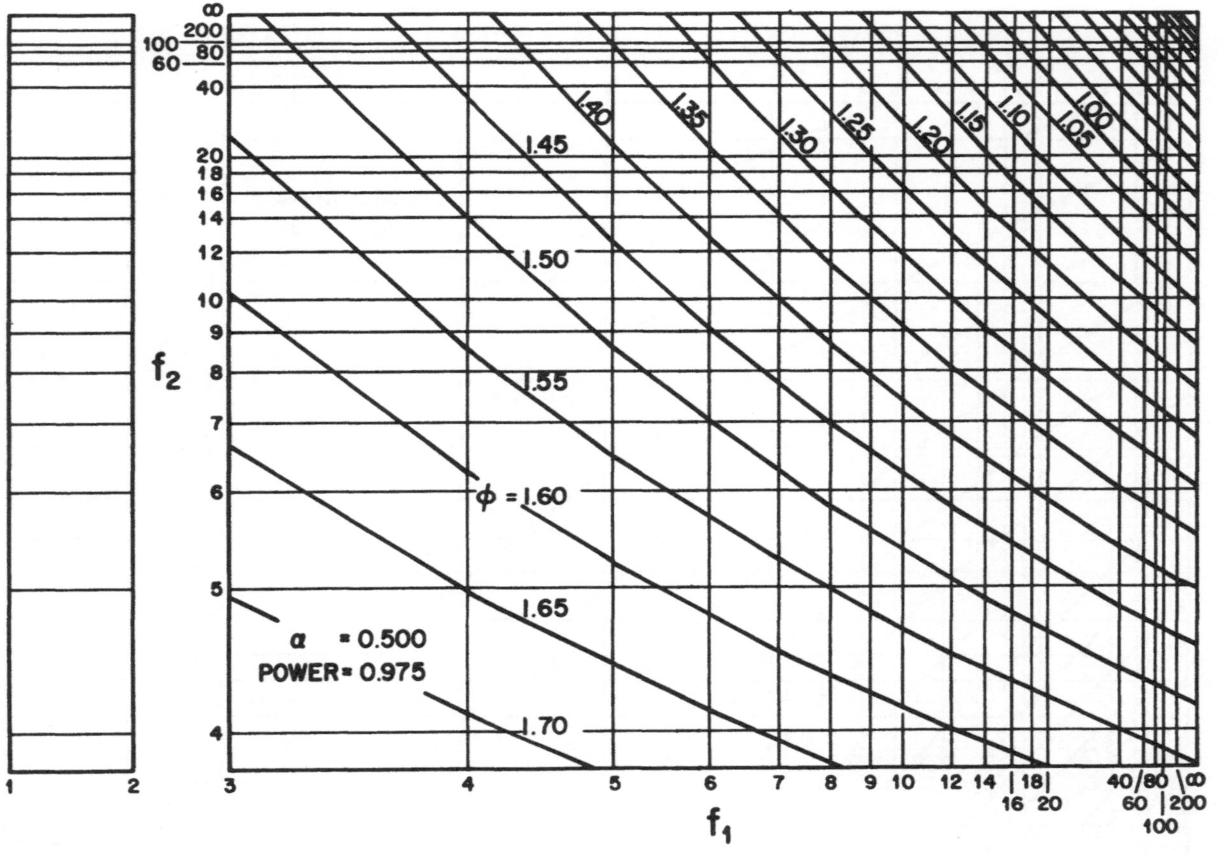

1.00
1.05
1.10
1.15
1.20
1.25
1.30
1.35
1.40
1.45
1.50
1.55
φ = 1.60
1.65
1.70
α = 0.500
POWER = 0.975
f2
f1

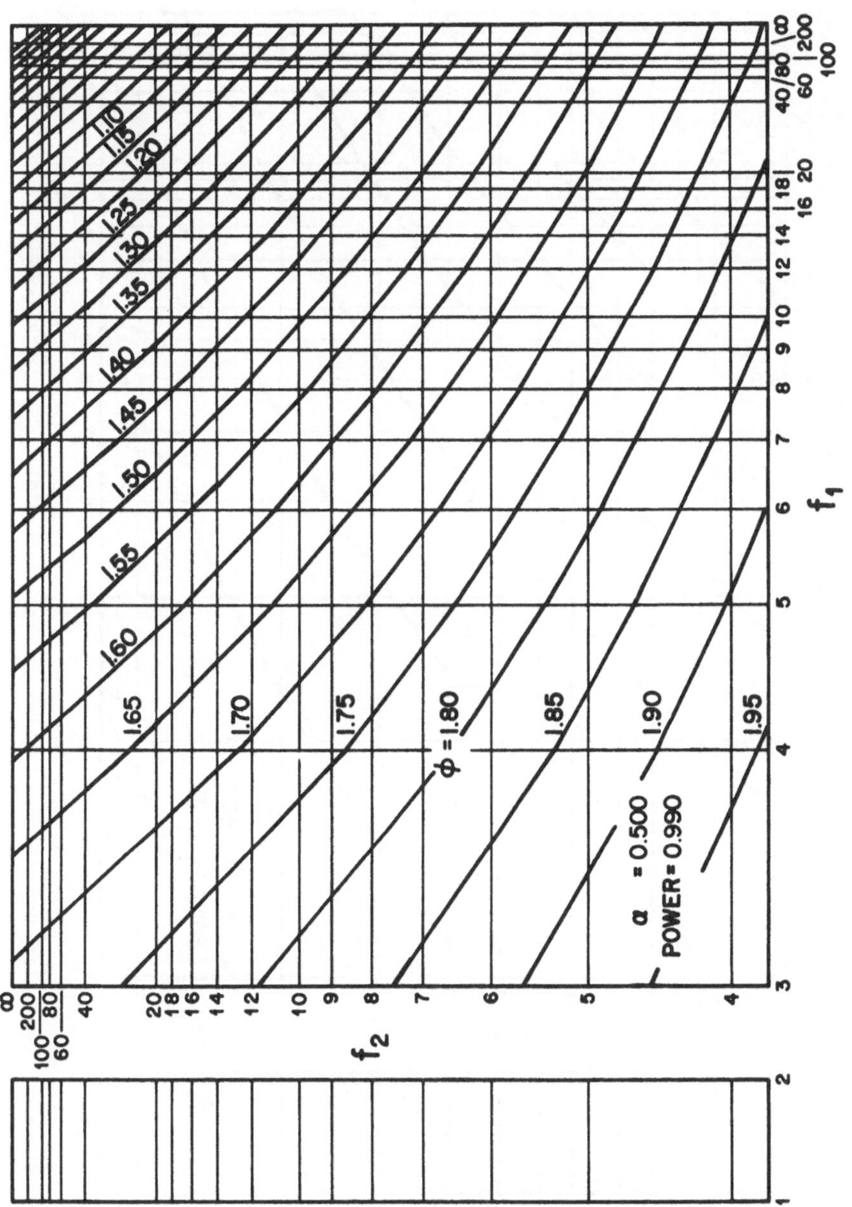
α = 0.500
POWER = 0.990
φ = 1.80
1.10
1.15
1.20
1.25
1.30
1.35
1.40
1.45
1.50
1.55
1.60
1.65
1.70
1.75
1.85
1.90
1.95
f_1
f_2
1 2 3 4 5 6 7 8 9 10 12 14 16 18 20 40 60 80 100 200 ∞
4 5 6 7 8 9 10 12 14 16 18 20 40 60 80 100 200 ∞

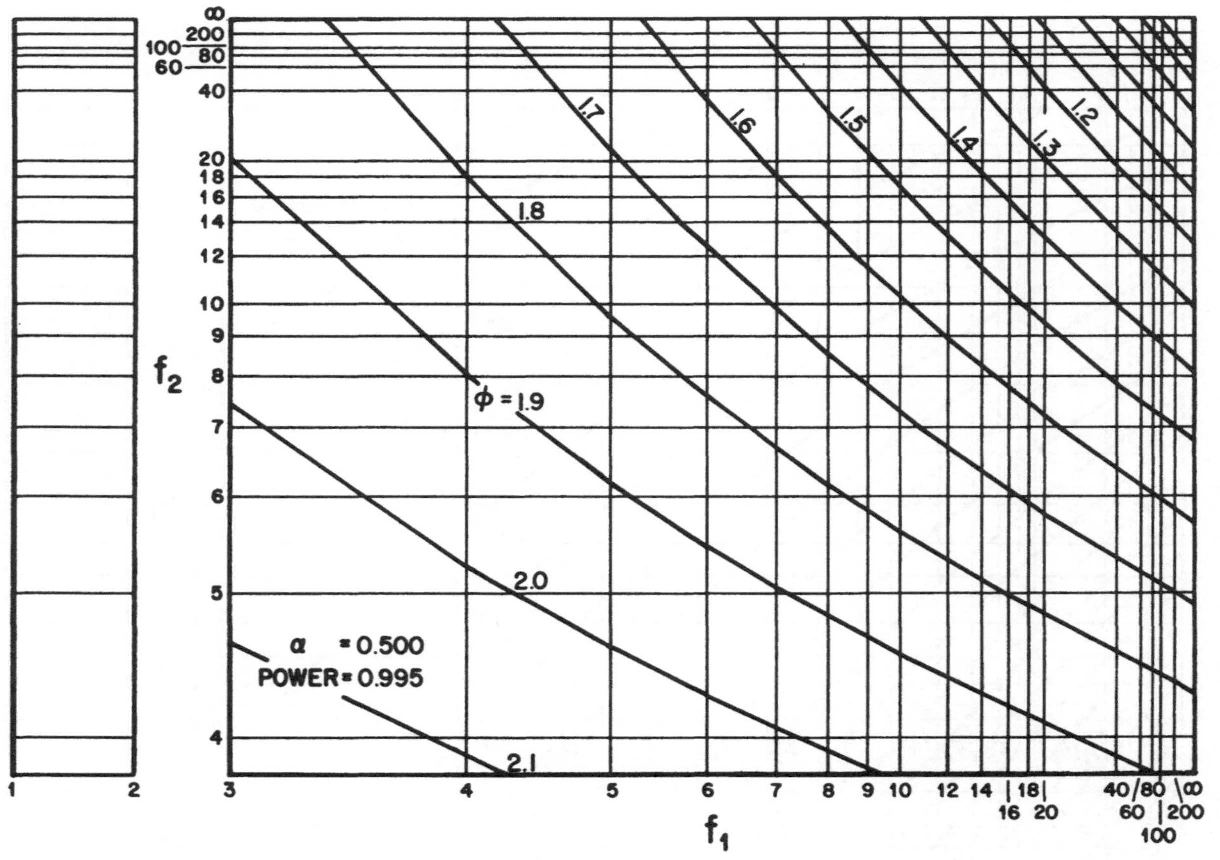

α = 0.500
POWER = 0.995
φ = 1.9
1.2
1.3
1.4
1.5
1.6
1.7
1.8
2.0
2.1
f_1
f_2

INDEX

9 780824 786007